Understanding the NEC4 Term Service Contract

Use of the NEC4 suite of contracts continues to grow and the new edition of *Understanding the NEC4 Term Service Contract* includes significant additional materials and changes since its original publication immediately after the initial release of the NEC4 contracts. Experienced authors and construction contracts specialists Kelvin Hughes and Patrick Waterhouse have added numerous practical experiences, case studies, lessons learned and guidance notes which were not available at the time of writing the original book.

Covering all the recent updates to the contract and written in plain English, *Understanding the NEC4 Term Service Contract* offers a practical guide to the use and management of the NEC4 Term Service Contract (TSC). The authors describe the full life of a contract, from the initial selection of options and contract formation through to the operations period and ultimately termination and dispute resolution. Although born of the same stable as the NEC4 construction contracts, the TSC is aimed at maintaining infrastructure and differs significantly from its siblings.

This is essential reading for anyone working with the contracts and takes the reader through the important provisions including communications, planning, early warnings, compensation events and payments. It is ideal for clients, contractors and their advisers describing how to deploy the contract successfully.

Kelvin Hughes is a leading authority on construction contracts and has had substantial theoretical and practical experience of all contracts, including NEC, JCT and FIDIC, over a 49-year working life, including four years as a full-time consultant to a major project management company drafting a suite of contracts for use on public projects in Qatar, and was also head of contracts/claims for another major project management company in Qatar. He has written eleven previous practical books on construction contracts.

Patrick Waterhouse is an adjudicator, mediator and expert determiner specialising in NEC, JCT and PFI contracts. He is a member of ten adjudicator nominating body panels. He has worked with NEC contracts since their first publication in the early 1990s. He is a chartered civil engineer, a chartered surveyor and a Fellow of the Chartered Institute of Arbitrators.

Understanding Construction

Understanding NEC3
Engineering and Construction Short Contract
Kelvin Hughes (2014)

Understanding The Building Regulations, 6th edition
Simon Polley (2014)

Understanding JCT Standard Building Contracts, 10th edition
David Chappell (2017)

Understanding NEC4
Term Service Contract
Kelvin Hughes and Patrick Waterhouse (2018)

Understanding the NEC4 ECC Contract
A Practical Handbook
Kelvin Hughes (2018)

Understanding the NEC4 Professional Service Contract
A Practical Handbook
Kelvin Hughes (2020)

Understanding FIDIC
The Rainbow Suite
Kelvin Hughes (2021)

Understanding the NEC4 ECC Contract
A Practical Handbook, 3rd edition
Kelvin Hughes (2024)

Understanding the NEC4 Term Service Contract
A Practical Handbook, 2nd edition
Kelvin Hughes and Patrick Waterhouse (2024)

Understanding the NEC4 Term Service Contract

A Practical Handbook

Second Edition

Kelvin Hughes and Patrick Waterhouse

Routledge
Taylor & Francis Group

LONDON AND NEW YORK

Designed cover image: Getty Images

Second edition published 2025
by Routledge
4 Park Square, Milton Park, Abingdon, Oxon, OX14 4RN

and by Routledge
605 Third Avenue, New York, NY 10158

Routledge is an imprint of the Taylor & Francis Group, an informa business

First edition published by Routledge 2020

British Library Cataloguing-in-Publication Data
A catalogue record for this book is available from the British Library

Library of Congress Cataloging-in-Publication Data
Names: Hughes, Kelvin (Engineering consultant), author. | Waterhouse, Patrick, author.
Title: Understanding the NEC4 term service contract : a practical handbook /
    Kelvin Hughes and Patrick Waterhouse.
Other titles: Understanding the new engineering contract third, term service contracts
Description: Second Edition. | Abingdon, Oxon [UK] ; New York, NY : Routledge, 2025. |
    Series: Understanding construction series | Includes bibliographical references and index.
Identifiers: LCCN 2024029327 (print) | LCCN 2024029328 (ebook) | ISBN 9781032733463
    (hardback) | ISBN 9781032726724 (paperback) | ISBN 9781003463771 (ebook)
Subjects: LCSH: Civil engineering contracts—Great Britain. | Construction contracts—
    Great Britain. | NEC Contracts.
Classification: LCC KD1641 .H8334 2024 (print) | LCC KD1641 (ebook) |
    DDC 343.4107/8624—dc23
LC record available at https://lccn.loc.gov/2024029327
LC ebook record available at https://lccn.loc.gov/2024029328

ISBN: 978-1-032-73346-3 (hbk)
ISBN: 978-1-032-72672-4 (pbk)
ISBN: 978-1-003-46377-1 (ebk)

DOI: 10.1201/9781003463771

Typeset in Times New Roman
by Apex CoVantage, LLC

Contents

Preface

Kelvin Hughes and Patrick Waterhouse, the joint authors of this book, have both been involved with the NEC family of contracts since their launch in the early 1990s.

In that time they have advised on numerous projects using the contracts, and between them have carried out well over 3,000 NEC-based training courses.

Kelvin was Secretary of the NEC Users Group from 1996 to 2006 providing support on behalf of the publishers to worldwide users of the contracts, and since that time has continued to provide training and consultancy advice on the full range of NEC contracts. He has also written several other books on the NEC and other contracts.

Patrick has worked as a contracts consultant and adjudicator advising his clients on contract issues and resolving disputes. His experience includes the full lifecycle of projects from initial conception through construction, operation and decommissioning.

During their involvement with the NEC contracts, the authors have always felt that there was a need for practical manuals on the contracts and their practical use, including worked examples and illustrations. For example, when considering the Term Service Contract how do you compile the Scope, how does the Price List work, what does a Task Order look like, etc.

The User Guides that accompany the contract were written by the contract drafters and whilst they are comprehensive and well written, the authors have always felt that they do not adequately act as a practical handbook. To the best knowledge of the authors, no other book on the Term Service Contract has been published, aside from the User Guides.

It is certainly not the authors' intention to criticise NEC, which they believe is an excellent family of contracts, well written by an extremely knowledgeable and professional Panel, many of whom they know personally and for whom they both hold a great admiration.

This book is intended to fill the gap where the User Guides leave off, and cover issues that may not have been considered or covered within the official publications.

It is intended to be of benefit to professionals who are actually using the contract, but also to students who need some awareness of the contract as part of their studies.

The Term Service contract was first launched in 2005 under the NEC3 family, the following editions and reprints to bring it up to the present day:

- First edition 2005
- Reprinted 2013
- NEC4 Term Service Contract launched June 2017.
- Amendments published in January 2019, October 2020 and January 2023.

References to clause numbers in the book relate to the latest January 2023 version, though the clause numbering is not totally dissimilar to the original June 2005 version.

The Term Service Contract brought some new concepts to the NEC family. Whilst retaining provisions such as Main and Secondary Options, Contract Data, early warnings, compensation events, etc the contract is based on a Contractor (who may be a Contractor in the normal sense, or a Consultant) providing a defined service at an Affected Property, which could be the Client's or someone else's premises, or a range of locations, rather than a Site, and over a fixed period of time.

Note that throughout the book the authors have made reference to the Term Service Contract using the initials "TSC".

As with the previous books, certain words are capitalised within the text for clarity or emphasis, e.g. Main Options and Secondary Options. Also, hopefully readers will not be confused in that, where referring to say "a contractor" lower case is used, but when referring to "the Contractor" (i.e. the one in the contract), the initial letter is capitalised.

As the authors have always viewed the NEC contracts as a manual of good practice as well as a contract, particularly in terms of the use of its Main and Secondary Options, risk management through the early warning system, clear requirement for a plan, and a disciplined procedure for change management, we have deliberately included within each chapter an overview of the subject area and how contracts deal with the issues, as well as detailing the NEC4 provisions.

Readers of the authors' other books on the NEC contracts, will see some duplication from those books when reading this one, particularly when giving general advice. This was deliberate, the intention being that each book should stand on its own, rather than one having to read the first book to properly understand the second book and so on.

Readers may note the absence of case law within the text of the book. This is a deliberate policy on our part, for three reasons:

(i) First, Kelvin and Patrick are not lawyers, Kelvin's background being in senior commercial positions with major building contractors and consultants, including working as an expert witness, and Patrick's being as an Adjudicator, Chartered Surveyor and Chartered Engineer, again with major civil engineering contractors and consultants, both having significant experience in NEC training, so it was felt, and readers may concur, that whilst we jointly have very significant experience with contracts and their administration, we were unqualified to quote, and to attempt a detailed commentary, on any case law.

(ii) Second, there has only been limited case law on the NEC contracts since they were first launched.

(iii) Third, and probably the most important, as contracts consultants with significant overseas experience of all contracts including the NEC, it was always our intention that the book should attract an international readership. NEC was always conceived as an international contract, so including UK case law would probably limit it to a British readership.

We have included within the text many examples, as we have found when running NEC training courses that delegates understand principles much better if they can be given illustrated and worked examples, and where numbers are involved using real calculations.

Our personal motivation for writing a book is to collate into one volume a significant quantity of material which we have gathered over the years and to share with others our substantial knowledge and experience of the contract.

Kelvin Hughes
Patrick Waterhouse
May 2024

Acknowledgements

I would like to extend my sincere thanks and gratitude to my wife Lesley, for the love, the time, the inspiration, and the support to fulfil my life's ambitions!

Also to Patrick, as we jointly update our previous book on the NEC4 Term Service Contract, it's a pleasure to know and work with him and to be able to combine our many years of knowledge and experience of using the NEC contracts!

This book is dedicated to my six grandchildren Grace, Emily, Thomas, Joshua, Ellie and Meadow, and I hope that in the years to come they will be proud of their granddad.

Kelvin Hughes
May 2024

I would like to thank all of those I have met in my professional life who have supported, helped and challenged me. Friends, clients and others all help me to continue learning every day. Their experiences have influenced my approach to this book as has my work with other NEC practitioners including Kelvin.

Away from work I am indebted to my family who continue to humour, encourage and support me.

Patrick Waterhouse
May 2024

Introduction to the NEC4 Term Service Contract

0.1 Introduction

NEC contracts including the Term Service Contract, are now being used substantially and by many clients in the public and private sectors in several countries, most clients having reported that the contract gives far greater control of time, cost and quality issues combined with greatly improved relationships between the contracting parties.

In June 2017 NEC4 was launched as a complete review and update of the whole NEC family, including the introduction of several new contracts, the Design Build and Operate Contract, the Professional Service Subcontract, the Alliance Contract, and the Dispute Resolution Service Contract. The Facilities Management Contract followed shortly afterwards, which we will discuss in Chapter 12.

While articles written about NEC tend to give prominence to many of the big "flagship" projects carried out using the contracts, of much more significance are the very many more modest size and value projects and services being procured. It is, also in this area that many of the problems have arisen in implementation of the contract.

The original objectives of the NEC contracts, and more specifically the Term Service Contract (TSC), which is the topic of this book, were to make improvements under three headings. Those three objectives remain to this day:

1. Flexibility

The contract should be able to be used:

- for any services provided to infrastructure containing any or all of the traditional disciplines such as civil engineering, building, electrical and mechanical work, and also process engineering.

Examples of the use of the TSC in the construction sector include:

- maintenance of highways in a particular area
- periodic inspection and reporting of bridge structures
- servicing and maintaining airport terminal buildings
- maintaining public parks and landscape areas
- maintaining heating, lighting and ventilation of buildings
- maintaining lifts in a group of hospitals
- maintenance of a canal and servicing the leisure facilities it affords

DOI: 10.1201/9781003463771-1

However, the TSC is also used, for non-construction sector services, for example:

- cleaning of streets in a town centre
- refuse collection and disposal
- provision of ambulance services for a group of hospitals
- the provision of data processing services by a computer systems company over a number of years
- the provision of security personnel for a building
- for Public Private Partnership enterprises perhaps alongside construction projects under the ECC, e.g. the construction of a new prison and subsequent running of the prison
- whether the Contractor has full, some or no design responsibility.

Design in TSC contracts tends to be limited by the nature of the work, but Task Orders often contain design obligations, and there are also often requirements for temporary access design, such as scaffolding and other temporary support systems.

- to provide all the normal current options for types of a service contract such as lump sum, cost-reimbursable or a target contract.
- to allocate risks to suit each particular project.

Often, other contracts are written with risks allocated by the contract drafters, and one has to be expert in contract drafting to amend the conditions to suit each specific contract:

- anywhere in the World.

Other contracts include country specific procedures and legislation and therefore either cannot be used or have to be amended to be used in other countries. To that end, FIDIC contracts are sometimes seen as the only forms of contract which can be used in a wide range of countries.

NEC contracts have been used for projects as widely diverse as airports, sports stadiums, water treatment works, housing projects, nuclear decommissioning in many parts of the works and even research projects in the Antarctic.

2.   *Clarity and simplicity*

- The contract is written in ordinary language and in the present tense.
- As far as possible NEC only uses words which are in common use so that it is easily understood, particularly where the user's first language is not English. Previously obscure words such as "whereinbeforesaid", "hereinafter" and "aforementioned" were commonplace in contracts. NEC contracts also have few sentences that contain more than 40 words and use bullet points to subdivide longer clauses.
- The number of clauses and the amount of text are also less than in most other standard forms of contract and there is an avoidance of cross referencing found in more traditional standard forms.
- It is also arranged in a format, which allows the user to gain familiarity with its contents, and required actions are defined precisely thereby reducing the likelihood of disputes.
- Finally, subjective words like "fair" and "reasonable" have been used as little as possible as they can lead to ambiguity, so more objective words and statements are used.

Some critics of NEC have commented that the "simple language" is actually a disadvantage as certain clauses may lack definition and there are certain recognised words that are commonly used in contracts. NEC has now been used for over 30 years, and still there is limited case law in existence with the contracts, and while adjudications are confidential and unreported, anecdotal evidence suggests that there does not appear to be any more adjudications with NEC contracts than any other, and probably less, which would tend to suggest that the criticism may be unfounded.

3. *Stimulus to good management*

This is perhaps the most important objective of the TSC in that every procedure has been designed so that its implementation should contribute to, rather than detract from, the effective management of the service. In order to be effective in this respect, contracts should motivate the parties to proactively want to manage the outcome of the contract, not just to react to situations. NEC intends the parties to be proactive and not reactive. It does also require the parties and those that represent them to have the necessary experience (sadly often lacking) and to be properly trained so that they understand how the NEC works.

The philosophy of NEC is founded on two principles:

- "Foresight applied collaboratively mitigates problems and shrinks risk"
- "Clear division of function and responsibility helps accountability and motivates people to play their part"

Examples of foresight within the TSC are the early warning and the compensation event procedures.

The early warning provision requires the Service Manager and the Contractor each to notify the other upon becoming aware of any matter which could have an impact on price or quality.

This provision can sometimes be misunderstood. A view held by many Service Managers is that an early warning is something that only the Contractor should give, and is an early notice of a "claim". This is an erroneous view as first early warnings should be given by either the Service Manager or the Contractor, whoever becomes aware of the matter first, the process being designed to allow the Service Manager and Contractor to share knowledge of a potential issue before it becomes a problem, and second, early warnings should be notified regardless of whose fault the problem is, it is about raising and resolving the problem, not compensating the affected party.

The compensation event provision requires the Contractor, within three weeks of being instructed, to submit a quotation to the Service Manager showing the cost effect of the event. The Service Manager then responds to the quotation within two weeks, enabling the matter to be properly resolved close to the time of the event rather than many months or even years later.

The plan is also an important management document with the contract clearly prescribing what the Contractor must include within its plan and requiring the Service Manager to "buy into it" by formally accepting (or not accepting) the plan. The Service Manager can use the plan to decide the most appropriate way to implement change.

In total, the TSC is designed to provide a modern method for Clients, Contractors, Service Managers and others to work collaboratively and to achieve their objectives more consistently than has been possible using other traditional forms of contract. People will be motivated to play their part in collaborative management if it is in their commercial and professional interest to do so.

Uncertainty about what is to be done and the inherent risks can often lead to disputes but the TSC clearly allocates risks and the collaborative approach will reduce those risks for all the parties so that uncertainty will not arise.

0.2   Essential differences between the TSC and other forms of contract

The TSC is part of a matrix of contracts, allowing all parties, whatever the project or service to be provided, to work under similar conditions. A full list can be found on the NEC website. The matrix includes the Term Service Short Contract, the Professional Service Contracts and the Facilities Management Contracts.

The TSC is a contract by which a Contractor provides a service to a Client for a period of time (a term) which is called the "Service Period". This period begins at the "starting date".

The TSC has been drafted using the same principles as in other contracts in the family of NEC documents. It uses similar procedures and wording as much as possible. The "Service Manager" manages the contract on behalf of the Client.

Flexibility of use: The TSC is not sector specific. It is used in a wide range of infrastructure maintenance scenarios including highways maintenance, accommodation facilities, healthcare, education and defence.

Flexibility of procurement: The TSC's Main and Secondary Options, allow it to be used for any procurement method, and can be bespoke to specific needs.

Early warning: The TSC contains express provisions requiring the Contractor and the Service Manager to notify and if required call an "early warning meeting" when either becomes aware of any matter which could affect price, timing or effectiveness of the service being provided.

Plan and Task Order programme: There is a clear and objective requirement for a detailed plan and Task Order programme, with method statements and regular updates which provides an essential tool for the parties to manage the project and to notify and manage the effect of any changes, problems, delays, etc.

Compensation events: This procedure requires the Contractor to price the "Defined Cost" effect of a change within three weeks and for the Service Manager to respond within two weeks. There is therefore a "rolling" final account with early settlement and no later "end of job" claims for delay and/or disruption. It is also more beneficial for the Contractor in terms of its cash flow as the Contractor is paid agreed sums rather than reduced "on account" payments, which are subject to later agreement and payment.

Disputes: The contract encourages better relationships and there is far less tendency for disputes because of its provisions. If a dispute should arise there are clear procedures as to how to deal with it using first the Senior Representatives, then adjudication and then the tribunal (either litigation or arbitration).

0.3   A "Service" not a "Project"

There are some important differences between "Providing the Work" (as in the ECC) and "Providing the Service" (as in the TSC).

The end result of an ECC contract is a project provided for the Client which did not previously exist. The TSC, however, makes provision for maintaining an existing asset, the end result of which is the same asset and in a similar condition as it was at the commencement of the Service Period.

During the Service Period, the Client will have had some benefit from the Contractor in the form of an asset which has been maintained, or some other specified benefit.

One consequence of these differences is that there can be no extension of time (delay to the Completion Date) in the TSC as in the ECC, because aside from Task Orders, the TSC does not use completion or completion dates. The TSC Service Period may of course be extended by agreement of the parties, in this case through Secondary Option X23.

The only exception to this is in the use of Task Orders. The discrete Task to be performed is described in a Task Order together with the period stated in the Task Order, for its performance. This period may be extended by means of a compensation event.

0.4 Range of use

The range of services covered by the TSC is extremely wide. It may extend from the maintenance of a nuclear power station to sweeping the streets.

The TSC is not a contract to provide a project. The principle of the TSC is based on providing a service, i.e. maintaining an existing condition for a period of time to permit the Client's continuing use of a facility.

It does not normally include the improvement of an existing condition of an asset – that would comprise a project. However, a modest amount of improving the condition of an asset - sometimes called "betterment" - may sometimes be sensibly included in a TSC.

For the purposes of the TSC, maintenance includes renewal and replacement of things which have become worn out, or otherwise reached the end of their useful lives.

There will usually be some infrastructure or asset (which may or may not belong to the Client) where the service is provided. The TSC refers to the areas under which the service is provided as the "Service Areas" and the part that is specifically affected by the work as the "Affected Property".

The "Scope" describes the service to be provided by the Contractor. It also includes full details of where and how it is to be provided, and any constraints placed upon the Contractor.

0.5 Contrast with the Professional Service Contract (PSC)

Like the Professional Service Contract (PSC), the TSC is a service contract designed for use in many sectors of commercial activity, and is not confined to use in the built environment industry.

The TSC differs from the PSC in that normally the service under the TSC will include physical work such as cleaning, painting and similar maintenance type operations but it is not restricted to such work. It may include routine inspection, maintenance and reporting operations as well as operations necessary to achieve defined performance criteria.

The TSC also differs from the PSC in that the latter is restricted to providing a specifically professional service. Also the main use of the PSC is for project work rather than maintenance type work, characteristic of most of the work which is envisaged to be done under the TSC.

More detailed differences between the TSC, the PSC and the FMC are covered within Chapter 12.

0.6 Mutual trust and co-operation

The first clause of all the NEC contracts has always required the Parties, and their agents, e.g. Service Manager, to act in a spirit of mutual trust and co-operation.

This has been slightly changed within the NEC4 contracts in that the former Clause 10.1 has now been split into two separate clauses, 10.1 and 10.2:

- Clause 10.1: *The Parties and the Service Manager shall act as stated in the contract.*
- Clause 10.2: *The Parties and the Service Manager act in a spirit of mutual trust and co-operation.*

If we examine the clauses, first, with Clause 10.1, is it necessary to state that the Parties and the Service Manager are required to act as stated in the contract? What difference does it make if that clause was absent, or deleted? Would they not have to act as stated in the contract?

Second, in Clause 10.2, what does it mean that the Parties and the Service Manager are required to act in a spirit of mutual trust and co-operation?

This second requirement has often been viewed with some confusion, and for those who have spent many years in the construction industry, with a degree of skepticism! Most practitioners state that their understanding of the clause is that the parties should be non-adversarial toward each other, acting in a collaborative way, and working for each other rather than against each other, and in reality that is what the clause requires.

However, the difficulty is, that if a party does not act in a spirit of mutual trust and cooperation what can another party do? The answer is that the clause is difficult to enforce. It is easy to imagine a failure of someone to co-operate, the second part of Clause 10.2. But case law in good faith obligations such as the first part of 10.2 shows that clear principles are difficult to identify. The Contractor is not contractually related to the Service Manager, so neither would be able to take action directly against the other for breach of contract. But the Client is liable for the faults of the Service Manager as this is a Client liability in Clause 80.1.

In effect, a clause requiring parties to act in a certain spirit will probably not, on its own, have any real effect. Within the TSC it is the clauses that follow within the contract which require early warnings, clearly detailed plans and programmes which are submitted for acceptance, and a structured change management process, that actually create and develop that level of mutual trust and co-operation rather than simply inserting a statement within the contract requiring the parties to do so.

0.7   Arrangement of the TSC

The TSC includes the following sections:

- Core clauses

  1. General
  2. The Contractor's main responsibilities
  3. Time
  4. Quality management
  5. Payment
  6. Compensation events
  7. Use of equipment, Plant and Materials
  8. Liabilities and insurance
  9. Termination

- Main Option clauses

  Option A: Priced contract with price list
  Option C: Target contract with price list
  Option E: Cost reimbursable contract

- Dispute resolution

  Option W1: Dispute resolution procedure (used unless the Housing, Grants, Construction and Regeneration Act 1996* applies)
  Option W2: Dispute resolution procedure (used in the United Kingdom, when the Housing Grants, Construction and Regeneration Act 1996* applies)

*See amendment regarding the Local Democracy, Economic Development and Construction Act 2009.

- Secondary Option clauses

  Option X1: Price adjustment for inflation (used only with Options A and C)
  Option X2: Changes in the law
  Option X3: Multiple currencies (used only with Option A)
  Option X4: Ultimate holding company guarantee
  Option X8: Undertakings to the *Client* or Others
  Option X10: Information modelling
  Option X11: Termination by the *Client* (not used with Option X19)
  Option X12: Multiparty collaboration (not used with Option X20)
  Option X13: Performance bond
  Option X17: Low service damages
  Option X18: Limitation of liability
  Option X19: Termination by either Party (not used with Option X11)
  Option X20: Key Performance Indicators (not used with Option X12)
  Option X21: Whole life cost
  Option X23: Extending the Service Period
  Option X24: The *accounting periods*
  Option X29: Climate change
  Option Y(UK)1: Project Bank Account
  Option Y(UK)2: The Housing Grants, Construction & Regeneration Act 1996
  Option Y(UK)3: Contracts (Rights of Third Parties) Act 1999
  Option Z: *Additional conditions of contract*

NB Options X5 to X7, X9, X14 to X16, X22 and X25 to X28 are not used

- Contract Data

Other documents to be used with the TSC include

- the Scope
- the Accepted Plan
- other documents resulting from choosing various Secondary Options, e.g. performance bond, undertakings to Others, etc.

Depending on the choice of Main Options the documents may also include:

- a Price List

0.8 Preparing the contract documents

Any contract under the TSC comprises:

- The core clauses
- One Main Option (Option A, C or E)
- The selected dispute resolution (Option W1 or W2).

- Secondary Options as desired
- Contract specific information (including Scope) in the Contract Data Part 1 provided by the Client
- Contract specific information in the Contract Data Part 2 provided by the Contractor
- The Price List

Where the Client invites potential contractors to tender, part two of the Contract Data is prepared in outline by the Client and completed by each tenderer. Similarly, the Price List is prepared by the Client and completed by the tenderer by pricing the items listed by the Client and possibly adding and pricing further items. The financial aspects of the tenders are then assessed by reviewing the information tendered in Contract Data Part 2 (particularly the fee percentage) and the rates and prices in the Price List.

0.9   The Price List

The Price List must be prepared for each contract. It will be seen from the above that the Price List consists of two kinds of entries:

- Lump sum items
- Remeasurable items

Where the Contractor is paid, say, an amount each month for the service it is providing, the description of the work covered by the sum of money is entered in the "Description" column, and the rate for each month is entered in the "Rate" column.

The total number of months is entered in the "Quantity" column. Thus the flexibility in payment methods is provided in the TSC by means of the different ways in which the Price List can be used.

Again to provide flexibility, the Price List can be used in a number of different ways:

 (i)   as a lump sum contract based on a specification and/or drawings;
 (ii) as a lump sum contract were the Client gives a list of work items and the Contractor is required to price each item as a lump sum. The total of the lump sums is the lump sum offer for the works;
(iii) as an admeasurement contract in which the Client lists the quantities and item descriptions and the tenderer enters a rate and extends the rate to the prices column. Here the itemisation is decided by the Client and is not subject to change except by compensation events. If the actual quantities carried out are different from those stated, they are corrected (see definition of Price for Service Provided to Date);
(iv) as an admeasurement contract in which the Client lists work items to be priced in the form of a lump sum, and remeasurable items as in (c) above; and
 (v) any of the above (a) to (d) but where the Contractor writes the Price List to the instructions of the Client in the invitation to tender or as a result of negotiation.

Lump sum contract based on specification and/or drawings

Here a brief description is given within the Price List, referenced to the Scope, which could be a specification based on deliverables and/or drawings, the Contractor pricing the work as a single lump sum, which could also be made up of a series of individual lump sums.

Procurement using specifications is still the way the largest number of service contracts, particularly those lower in value, are procured. The lump sum contract (Option A) has always been the most commonly selected alternative under the NEC contracts, although the target contract in Option C is also popular.

The basis of a lump sum contract is normally where the Client knows exactly what it wants, and is able to clearly define it through the Scope, which could comprise specifications, the latter possibly being a performance specification.

When the Contractor has priced the Price List, the lump sum for each item will be the Price to be paid by the Client to the Contractor in the assessment following completion of that item.

This is therefore effectively a stage payment contract and as payment is linked to completion of items, administration of payment aspects is therefore fairly simple and transparent.

It is important to recognise that the Price List has two primary functions:

1. It shows how the tendering Contractors have built up their price and can therefore be used as part of the tender assessment process.
2. It is used to calculate the Price for Service Provided to Date in Main Option A.

There are certain advantages to the Client using the lump sum alternative:

- There is no requirement for a detailed pricing document to be prepared by the Client for tendering Contractors to price. This will save the Client time and money at the pre-tender stage.
- The Contractor holds the risk of inaccuracies in the quantities within the lump sum price. However, it must be stressed that the Client holds the risk of inaccuracies and omissions within the Scope itself.
- The assessment of the Price for Service Provided to Date is easier and quicker than with the other alternatives. There is no requirement to remeasure the work and apply quantities to rates and prices to calculate the amount due to the Contractor.
- Plan and Price List preparation are linked as integrated activities which would tend to lead to a more comprehensive tender.
- Payment is linked to completion of an item or group of items related back to the programme, so cash flow requirements for both parties are more visible.
- The assessment of the effects of compensation events is related back to the Price List. Any change in resources or methods associated with an activity can be compared with those stated in the Accepted Plan before the compensation event occurred.

The disadvantages to the Client are:

- As there is no specific document that all the tendering Contractors price, it is difficult to assess tenders on a "like-for-like" basis. Some Clients cite that as an advantage, saying that they do not require the facility to check tenders on a line-by-line basis anyway!

  It is not uncommon, to assist the assessment of tenders, for Clients to issue templates for the Contractors to price covering the main elements as part of the tender with the tendering Contractors then inserting the items on the Price List under each element. In that way, each element will have been priced on a "like-for-like" basis, but the operations within each element may have been described and priced differently by each tenderer.

  If a template is used, Clients should be wary of them being too prescriptive and almost being seen by tenderers as a bill of quantities.

- The Client holds the risk of clearly defining the Scope in order that the Contractor can prepare the Price List.

Lump sum contract where the Client gives a list of service items and the Contractor is required to price each item as a lump sum. The total of the lump sums is the lump sum offer for the works

This is similar to pricing alternative (a), but in this case the Client itemises the various elements of the work within the Price List referenced to the Scope, the Contractor pricing the work as a series of individual lump sums.

(i) Admeasurement type contract in which the Client lists the quantities and item descriptions and the tenderer enters a rate and extends the rate to the prices column. Here the itemisation is decided by the Client and is not subject to change except by compensation events. If there is an error in that the actual quantities carried out are different from those stated, the Contractor is paid for the actual amount of service provided.

This alternative is a remeasurement contract with the Price List based on a pricing document prepared by the Client with the Contractor pricing the items in the Price List, including matters which are at its risk. The Contractor is then paid for the quantity completed to date at the rates and Prices in the Price List.

This alternative is normally used where the Client knows what it wants and is able to clearly define it through the Scope, and measure it within the Price List, but there are likely to be changes in the quantities, that may or may not be considered as compensation events.

Confusion can sometimes arise as to whether the tendering Contractors should price the Price List or the Scope. The Contractor Provides the Service in accordance with the Scope (Clause 20.1), therefore it is assumed that it programmes the Service in accordance with the Scope.

Information in the Price List is not Scope (Clause 55.1), it is purely for the purposes of payment. It does not tell the Contractor what it has to do, but it must clearly be a fair representation of the Scope to allow the Contractor to accurately price the Service.

Although a Contractor will often use the quantities within the Price List as a guide to calculate time scales, the Price List is used as a basis for inviting and assessing tenders, so it is a pricing and payment document, i.e. it deals with money. It is reasonable for the Contractor to presume that the quantities within the Price List are a fair representation of what is included in the Scope.

If there is an ambiguity or inconsistency within the Scope, either the Contractor or the Service Manager notifies the other as soon as either becomes aware that it exists and the Service Manager instructs a change to correct it. This may then be addressed as a compensation event under Clause 60.1 owing to any change to the Scope, with Clause 63.11 stating that where an instruction is given to change the Scope to resolve an ambiguity or inconsistency, it is assessed as if the Prices and the Task Completion Dates were for the interpretation most favourable to the Party which did not provide the Scope.

Clearly, if the Contractor notices the error during the tender period then it should raise it as a query, then the matter can be dealt with and all tenderers informed, however, while the Contractor should notify if and when it finds an error, it is not under any obligation to look for the error when tendering.

(ii) Admeasurement contract in which the Client lists work items to be priced in the form of a lump sum, and remeasurable items as in (c) above.

This is similar to pricing alternative (c), again this would create a remeasurement contract based on a pricing document within the Price List prepared by the Client with the Contractor pricing the items in the Price List, including matters which are at his risk, but there would be lump sum items as well as measured works. Again, the Contractor is then paid for the quantity completed to date at the rates and Prices in the Price List.

(iii)   Schedule of rates

This alternative is again a remeasurement contract based on the Price List prepared by the Client with the Contractor pricing the items in the Price List including matters which are at his risk, but for this alternative, there are either no quantities or estimated quantities so that the Client is able to compare tenders. The Contractor is then paid for the quantity completed to date at the rates and Prices in the Price List.

This alternative is often used for responsive maintenance and repair contracts, where the quantities are not known and cannot even be estimated.

0.10   The Main and Secondary Options

The three Main Options A, C and E enable Clients to select a procurement strategy and payment mechanism most appropriate to the service and the various risks involved. Chapter 5 includes more details on payment mechanisms and requirements under each Main Option.

Essentially, the Main Options differ in the way the Contractor is paid. There is an increasing use of target cost contracts which has been encouraged by the increasing use of partnering arrangements, and the better sharing of risk.

Once the procurement strategy has been decided, the Main and Secondary Options can be selected to suit that strategy.

The Main Options

Option A: Priced contract with price list
Option C: Target contract with price list
Option E: Cost reimbursable contract

- Option A is a priced contract in which the risks of being able to carry out the service at the agreed prices are largely borne by the Contractor.
- Option C is a target contract in which the Client and Contractor share the financial risks in a pre-agreed proportion.
- Option E is a cost-reimbursable contract in which the financial risks of being able to carry out the service are largely borne by the Client.

Option A: Priced contract with price list

Option A is normally used where the Client knows exactly what it wants, and is able to clearly define it through the Scope, which would mainly comprise specifications, so the Contractor can commit itself to a lump sum price and price the Price List which will identify its payments.

Option A is therefore effectively a stage payment contract as payment is linked to completion of items in the Price List plus any work where there are quantities stated, the Contractor must plan and carry out its work effectively with the cash flow requirements for both Parties being clearly visible. Administration of payment aspects is therefore fairly simple.

The price for each item is in effect a lump sum for that activity and must include for everything necessary to complete the activity. The sum of the tendered lump sums for each of the items is the tendered prices (the total of the Prices) for the whole of the service.

Option C: Target contract with price list

Option C is a target contract based on a Price List.

In Option C, the Price List acts as a guide when assessing tenders, but payments are based on Defined Cost incurred, not completion of items in the Price List.

Target contracts are a development of cost-reimbursable contracts and are a combination of a lump sum price and open book cost-reimbursable payments.

The advantage of target contracts is that the parties initially have certainty of price, with the Contractor being incentivised to make cost savings for the benefit of the Client and itself.

This Option is normally used where the Client knows what it wants and is able to clearly define it through the Scope, so that the Contractor can price the Price List, but sees a benefit in sharing risk and opportunity with the Contractor thereby encouraging collaboration. Financial risks are shared, proportionally between the Client and the Contractor through the Contractor's share percentages.

Some wrongly call Option C, the "Partnering Option" although it does lend itself to the principles of partnering, collaborative working and associated risk/opportunity sharing. Coincidentally, there is a multiparty collaboration Option (Option X12). This Option is described in detail later within this Chapter.

The Contractor tenders a price based on a Price List in the same way as it would under an Option A contract. This price, when accepted is then referred to as the "target". The Contractor also tenders its percentage for the Fee. The original target is the total of the Prices at the Contract Date.

- The target price includes the Contractor's estimate of Defined Cost plus other costs, overheads and profit to be covered by the Fee.
- The Contractor tenders the Fee in terms of a percentage to be applied to Defined Cost.
- During the course of the contract, the Contractor is paid Defined Cost plus the Fee.
- The target is adjusted for compensation events and also for inflation (if Option X1 is used)
- After the end of the Service Period, and at defined points during the Service Period, the Service Manager assesses the Contractor's share in accordance with Clause 54, which, it has to be said, is at best a confusing clause, though the User Guides clarify the clause! The Contractor then pays or is paid its share of the difference between the final total of the Prices and the final Price for Service Provided to Date according to a formula stated in the Contract Data. This motivates the Contractor to decrease costs. Many refer to this sharing of risk and opportunity as "pain and gain". It often comes as a surprise to NEC users that the terms "pain and gain" do not appear anywhere within the NEC contracts, nor does the term "target price"!

Under Options C and E, the Contractor is required to submit forecasts at the intervals stated in the Contract Data of the forecast total Defined Cost (Clause 20.4), which advises the Client of the likely outturn cost.

Option E: Cost-reimbursable contract

Option E is a cost-reimbursable contract with the Contractor being reimbursed Defined Cost plus the Fee.

It should be used:

- where the scope of the service is uncertain
- where extreme flexibility is required
- where a high level of Client involvement is envisaged
- for emergency work
- where trials or work of an experimental nature are carried out

A cost-reimbursable contract should be used where the definition of the service to be provided is inadequate even as a basis for a target price and yet an early start is required. In such circumstances the Contractor cannot be expected to take risks. It carries minimum risk and is reimbursed its Defined Cost plus Fee, subject only to a number of constraints designed to motivate efficient working.

A criticism of cost-reimbursable contracts such as Option E is that it gives the Contractor very little incentive to reduce costs. However, a cost-reimbursable contract should be used where the definition of the work to be done is inadequate for the Contractor to price and yet an early start is required. Payments are subject to any deduction for Disallowed Costs.

Under Option E of the TSC, the Contractor is reimbursed its Defined Cost plus a Fee, basically covering costs incurred within the Service Areas and profit. This Fee is calculated by applying the fee percentage, given at tender by the Contractor in Contract Data Part 2, to appropriate Defined Cost. Although Option E is a cost-reimbursable contract, there is an obligation on the Contractor to provide a regular forecast of the remaining Defined Cost which advises the Client of the potential outturn cost.

Another criticism of Option E is that the tendering Contractors do not actually price the service, they just price their fee percentage. In addition tenderers may be required to notionally agree to a cost plan. This makes it difficult to assess cost-reimbursable tenders, but again the basis of cost-reimbursable contracts is that the Client bears most of the risk.

Secondary Options

Option X1: Price adjustment for inflation (used only with Options A and C)

This Option can only be used with Options A and C. It is particularly suitable for service contracts where inflation is unpredictable, or where the Service Period is lengthy, say ten years, and thus where inflation may represent a high risk.

The Client should make the decision at the time of preparing the tender documents as to whether inflation for the duration of the contract is to be:

- The Contractor's risk: in which case it should not select Option X1
- The Client's risk: in which case, it should select Option X1

The default within the TSC is that the contract is "fixed price" in terms of inflation, i.e. the Contractor has priced the work to include any inflation it may encounter during the Service Period. If Option X1 is chosen, the Prices are adjusted for inflation as the work progresses, by means of a formula. Inclusion of this Option effectively transfers the risk of inflation to the Client.

The Option uses the formula method utilising published indices for rises in costs for various elements of the service.

- When used with Option A, the formula is applied to the rates and Prices in the Price List and the rates in the Contract Data for people and Equipment. This happens at each inflation adjustment date. That date is identified in the Contract Data and is usually annually but could be at any other frequency. This has the effect of adjusting the Price for Service Provided to Date which is derived from the Price List and, where there are compensation events and/or tasks, the Contract Data rates.

- When used with main Option C, the formula is applied to rates and Prices in the Price List and the rates in the Contract Data for people and Equipment. This happens at each inflation adjustment date. That date is identified in the Contract Data and is usually annually but could be at any other frequency. This has the effect of increasing both the Price for Service Provided to Date (the regular payments) and the Prices (the target).

The key components of the formula are:

- the "Base Date Index" (B) used for the first inflation adjustment date is the latest index available before the Base Date. The Base Date Index used for subsequent inflation adjustment dates is the index at the previous inflation adjustment date.
- the "Latest Index" (L) is the latest available index at the inflation adjustment date.
- the "Price Adjustment Factor" is the total of the products of each of the proportions stated in the Contract Data multiplied by $(L - B)/B$ for the index linked to it.

OPTION A

Under Option A, on each inflation adjustment date until the end of the Service Period, the amount due includes an amount for price adjustment, which is the sum of:

- the rates and Prices in the Price List; and
- the rates in the Contract Data for people and Equipment.

are changed by multiplying the rate or Price by $(1 + PAF)$. In calculating the Price for Service Provided to Date, the changed rates and Prices in the Price List are used for all works and services carried out from the inflation adjustment date until the next inflation adjustment date.

As an example, for simplicity ignoring correcting amounts:

The change in the Price for Service Provided to Date = £150,000

The Base Date Index (B) = 280.0

The Latest Index (L) = 295.5

The Price Adjustment Factor is therefore $(L - B)/B$

$= (295.4{-}280.0)/280.0$

$= 0.055$

Inflation since the base date is therefore 5.5%

The amount due is therefore £150,000 x 0.055 = £8,250.00

OPTION C

Under Option C, on each inflation adjustment date until the end of the Service Period, the amount due includes an amount for price adjustment which is the sum of:

- the rates and Prices in the Price List; and
- the rates in the Contract Data for people and Equipment.

are changed by multiplying the rate or Price by (1 + PAF). In calculating the Contractor's share, the changed rates and Prices in the Price List are used for all works and services carried out from the inflation adjustment date until the next inflation adjustment date. In calculating the Price for Service Provided to Date, the changed rates in the Contract Data are used for all works and services carried out from the inflation adjustment date until the next inflation adjustment date.

OPTION X2: CHANGES IN THE LAW

This Option transfers the risk of changes in the law after the Contract Date, from the Contractor to the Client. This risk may be high in some countries or where there is a long Service Period.

As with Option X1, the default is that the contract is "fixed price" in terms of changes in the law, i.e. the Contractor has priced the work to include any changes in the law it may encounter during the period of the contract.

If Option X2 is chosen, and a change in the law occurs after the Contract Date the Service Manager notifies the Contractor of a compensation event. The Prices may be increased or reduced in addition to providing for any delay to Task Completion Dates.

Note that Option X2 refers to a change in the law of the country in which the Affected Property is located, so for example, a change in the law in another country where goods are being fabricated for delivery to the Affected Property will not be a compensation event. X2 provides for a compensation event for a change in law that occurs after the Contract Date, even if it was known about prior to the Contract Date.

OPTION X3: MULTIPLE CURRENCIES (USED ONLY WITH OPTION A)

The currency of the contract is stated in Contract Data Part 1. However, Option X3 provides for items or activities to be paid in an alternative currency, the items or activities, the currency and the total maximum payment in this currency to be listed in the Contract Data, beyond which payments are made in the currency of the contract.

The exchange rates, their source and date of publication are also referred to in the Contract Data.

OPTION X4: ULTIMATE HOLDING COMPANY GUARANTEE

This form of guarantee (referred to in other contracts as a parent company guarantee) is given by a parent company (or holding company) to guarantee the proper performance of a contract by one of its subsidiaries (the Contractor), who while having limited financial resources itself, may be owned by a larger financially sound parent company.

In most cases it is the ultimate holding company that provides the guarantee, but sometimes, particularly when the ultimate holding company is in another country, the guarantor company may just be a company further up the chain within the group, perhaps the national parent who has sufficient assets to provide the required guarantee. If a company other than the ultimate holding company is proposed, the guarantor may not be accepted by the Service Manager as its commercial position is not strong enough to carry the guarantee.

If such a guarantee is required, it may either be provided by the Contract Date or within four weeks of the Contract Date.

Ultimate holding company guarantees are sometimes used as an alternative to a performance bond (Option X13). Sometimes both are demanded.

A holding company guarantee is cheaper than a performance bond, as the Contractor will normally just charge an administration fee rather than the case of performance bond, where the

Contractor is actually paying a premium for an insurance policy by an external provider, but it may give less certainty of redress because it is not supplied by an independent third party so it is dependent on the survival, and the ability to pay, of the ultimate holding company.

However, while accepting less independence, ultimate holding company guarantees for the proper performance of the contract can be more advantageous than bonds. Rather than receiving a fixed amount in compensation, the parent company is normally obliged to either complete the works in accordance with the Contractor's original obligations on behalf of its subsidiary company, or fund the completion of the contract by others, so effectively completion is guaranteed.

In the case of performance bonds the "guarantee" is that a sum of money is available to at least partly compensate the Client in the event of the Contractor's default.

Because the financial strength of the parent company may be linked to that of the Contractor, an ultimate holding company guarantee will be acceptable only if the parent company (or holding company) is financially strong and its financial resources are largely independent of those of the Contractor. Obviously, if the insolvency is not limited to the Contractor but also relate to the parent group an ultimate holding company guarantee will be virtually useless, save for any entitlement under insolvency law.

OPTION X8: UNDERTAKINGS TO THE CLIENT OR OTHERS

This Option provides for the Contractor to give undertakings to Others as stated in the Contract Data and, if required, in the form stated in the Scope. This may also include undertakings between a Subcontractor and Others if required by the Client. Typically such documents are often referred to as collateral warranties.

The Client prepares the undertakings and sends them to the Contractor for signature, the Contractor signs or arranges for the Subcontractor to sign them, and returns them to the Client within 3 weeks. For these provisions to be enforceable, the final wording of the undertakings must be included in the Scope at the Contract Date.

OPTION X10: INFORMATION MODELLING

This is a new Option which provides for Building Information Modelling.
Some defined terms should be considered in reviewing Option X10:

- The Information Execution Plan is submitted by the Contractor to the Service Manager who then accepts or does not accept within two weeks of receiving it. The Information Execution Plan is defined as the *information execution plan* (identified within the Contract Data), or the latest submitted by the Contractor and accepted by the Service Manager.
- Project Information is provided by the Contractor and used to create or change the Information Model.
- The Information Model is the electronic integration of the Project Information, and other information provided by the Client and other Information Providers. The Client is liable for any fault or error in the Information Model, unless there is a Defect in the Project Information provided by the Contractor. The Contactor is required to provide insurance against claims made in respect of failure to provide the Project Information using reasonable skill and care.
- The Information Model Requirements are the requirements identified in the Scope for creating or changing the Information Model.

There is also provision within the Option for early warnings where something could affect the Information Model, and for the Contractor to include within a quotation for a compensation event where the Information Execution Plan is altered by a compensation event.

OPTION X11: TERMINATION BY THE CLIENT (NOT USED WITH OPTION X19)

This Option provides for the Client to terminate the Contractor's obligation to Provide the Service for a reason not stated in the Termination Table, by notifying the Service Manager and Contractor.

If the Client does terminate, it may complete the service and use any Plant and Materials provide by the Contractor (Procedure P1), and the Contractor leaves the Service Areas and removes the Equipment (P4).

In respect of payments, the Contractor is entitled to payment amount A1:

Amount A1
• an amount due as for normal payments
• the Defined Cost for Plant and Materials which have been delivered and retained by the Client or which the Client owns and which the Contractor has to accept delivery
• other Defined Cost in reasonable expectation of completing the whole of the service (such as long-term supply contracts for consumables)
• any amounts retained by the Client

Amount A2
• The forecast cost of removing Equipment

Amount A4
• for Options A and C, the fee percentage applied to the difference between the original total of the Prices and the Price for Service Provided to Date
• for Option E, the fee percentage applied to the difference between the first forecast of the Defined Cost for the service and the original total of the Prices and the Price for Service Provided to Date less the Fee

OPTION X12: MULTIPARTY COLLABORATION (NOT USED WITH OPTION X20)

This is the Option which, if included in NEC contracts, creates additional obligations for the Partners listed in the Contract Data.

The Option enables a multi-party partnering agreement to be implemented. In this case Option X12 is used as a Secondary Option common to the contract which each party has with the body which is paying for the work. The parties together make up the partnering team.

It must be stressed that no legal entity is created between the Partners, so it is not a formal "partnership" as such.

Some definitions need to be explained:

 (i) The Partners are those named in the Schedule of Partners
 (ii) An Own Contract is a contract between two Partners, which includes Option X12
(iii) The Core Group comprises the Partners listed in the Schedule of Core Group Members.
(iv) Partnering Information is information which specifies how the Partners work together.
 (v) A Key Performance Indicator is an aspect of performance for which a target is attached in the Schedule of Partners.

Each Partner, represented by a single individual is required to work with the other Partners in accordance with the Partnering Information to achieve the Promoter's objective stated in the Contract Data and the objectives of every other Partner. The Promoter is named in the Contract Data.

The Core Group, led by the Promoter's Representative, its members selected by the Partners, acts and takes decisions on behalf of the Partners. The Core Group also keeps up to date, a Schedule of Core Group Members and a Schedule of Partners.

The Partners are required to work together, using common information systems, and a Partner may ask another Partner to provide information which it needs to carry out the work in its Own Contract. Each Partner gives an early warning to the other Partners when it becomes aware of anything that could affect another Partner's objectives.

The Core Group may give an instruction to the Partners to change the Partnering Information, which is a compensation event.

The Core Group also maintain a timetable showing the timing of the Partners' contributions. If the Contractor needs to change its plan to comply with the timetable then it is a compensation event.

Each Partner also gives advice, information and opinion to the Core Group where required.

Each Partner must also notify the Core Group before subcontracting any work, though it does not say that the Core Group is required to respond to the notification.

Finally Option X12 provides for Key Performance Indicators (KPI) with amounts paid as stated in the Schedule of Partners. The Promoter may add a KPI to the Schedule of Partners but cannot delete or reduce a payment.

OPTION X13: PERFORMANCE BOND

A performance bond is an arrangement whereby the performance of a contracting party (the Principal) is backed by a third party (the Surety), which could be a bank, insurance company or other financial institution, that should the Principal fail in its obligations under the contract, normally owing to the Principal's insolvency, the Surety will pay an amount of money, up to a pre-agreed maximum, to the other contracting party (the Beneficiary).

While bonds are not always linked to insolvency, contracts normally give remedies in the event of default, e.g. defects provisions, withholding of payment, etc. It is where the defaulter is not able to provide a remedy as it is insolvent that bonds normally provide the remedy.

This bond may be between any two contracting parties, Client and Contractor, Contractor and Subcontractor, or any other contractual relationship.

Note that as these are conditional bonds, it is normally a condition that default has to have occurred, there is no other means of remedy or recompense and the amount claimed is directly related to the default, before payment can be made to the beneficiary.

Clearly, if a bank provides a bond there may be some implications with regard to further credit as a safeguard should the bond be called, the bank also pursuing any amount paid from their customer, whereas with an insurance company the risk is accepted through the policy.

If a performance bond is required from the Contractor it should be provided by the Contract Date, or within four weeks of the Contract Date. The amount of the bond, often 10 per cent of the contract value must be stated in Contract Data Part 1 and the form of the bond must be in the form stated in the Scope. In the event of the default by the Contractor the Client is not paid the full amount of the bond, but the amount of the bond is the maximum amount available to meet the Client's costs incurred as a direct result of the default.

Normally, the value of the performance bond does not reduce, but they should have an expiry date, which could be completion of the service, or may even, in exceptional cases, include the six-, ten- or twelve-year limitation period following completion of the works to cover any liability for potential latent defects, the expiry date being defined within the Scope.

It is worth mentioning that, unless the bond is an on-demand bond, payment is often very difficult and subject to a number of conditions for example clear evidence of the Contractor's inability to perform its obligations, and also unequivocal evidence of the Client's loss as a direct result of the Contractor's default, which in truth may not be fully known for some time after the Contractor's insolvency. Clearly the expiry date for the bond has to be kept in mind if it takes some time to ascertain the Client's actual loss as the right to claim may have expired.

The bank or insurer which provides the performance bond must be accepted by the Service Manager.

It is important to note the difference between an insurance policy and a bond in that an insurance policy is a contract between TWO parties, the insured and the insurer, the insurer guaranteeing that it will pay a third party, who may not be specifically named, should an event occur, while a bond is a contract between THREE parties, the Contractor, the Client and the Surety, all of whom are named within the agreement, no other party being able to benefit from the bond.

A performance bond will not of itself ensure that contracts are carried out efficiently and to time, but it will be one of the number of commercial pressures on the Contractor to perform well. A performance bond can provide some compensation if the Contractor defaults on its obligations. The cost of the performance bond will normally be governed by:

(i) The technical ability of the Contractor
(ii) The usual type of work undertaken by the Contractor
(iii) The scale of the service to be provided
(iv) Contracts already bonded for the Contractor
(v) The overall management of the Contractor's business
(vi) The financial stability of the Contractor

There are other types of bond available:

(i) Advanced Payment Bond

This covers the Contractor's repayment of an advanced payment from the Client to the Contractor. The TSC does not contain provisions for such a bond, but other NEC contracts do. For one to be included, additional conditions would need to be incorporated.

(ii) Payment Bond

This covers the Client's duty to pay the Contractor, or the Contractor's duty to pay its Subcontractors.

(iii) Bid or Tender Bond

Very often on contracts with a long lead in time, a bond may be required to ensure that the successful tenderer will not withdraw its tender and is able to proceed when required, in terms of its other commitments and its solvency.

In a similar vein, a scenario may be where a Contractor prices a tender based on a Subcontractor's price and finds upon acceptance of the tender that the Subcontractor's offer is no longer open for acceptance, and the Contractor may find itself having to place orders with another Subcontractor for a higher price.

It may be worthwhile in this event to bond the Subcontractor so that in the event that the acceptance period expires, the Contractor can recover monies through the bond.

Unconditional or on Demand Bonds   "Unconditional" or "On Demand Bonds" can be redeemed by the beneficiary, whether or not there has been a default within the contract and whether or not there has actually been a loss at all!

Most Clients will refrain from calling in bonds unnecessarily, but the entitlement remains should they wish to.

The "knock on" effect of calling in these bonds without due cause is that Contractors will price for that possibility within their tenders, as someone has to pay, and the result will be escalating prices. Insurers are naturally very reluctant to accept on demand bonds and the market often lacks sufficient liquidity for such arrangements.

Creation of Bonds   Bonds have to be made in writing and will be signed and sealed by the contracting parties.

The duration and financial limits of the Bond MUST be clarified and stated within the deed.

Release and cancellation of Bonds   The deed executed will normally specify the course of action to be taken in the event of failure of the Principal leading to the calling in of the Bond by the Beneficiary.

Sometimes the Beneficiary has to give a form of notice to the Principal or it may have to sue for non-performance.

Only the Beneficiary can cancel the Bond, the Principal issuing a request to them, once it has "performed".

OPTION X17: LOW SERVICE DAMAGES

In the event that the Contractor produces a defective service which does not meet the level stated in the service level table, the Client has three options:

(i)   The Contractor corrects the Defect (Clause 43.2)
(ii) If the Contractor does not correct the Defect, the Service Manager assesses the cost to the Client of having the Defect corrected by other people and the Contractor pays this amount (Clause 43.3)
(iii) The Client can accept the Defect and a quotation from the Contractor for reduced Prices and/or an earlier Task Completion Date (Clause 44)

The Client can also, as an alternative, recover low performance damages under Option X17 if it has been selected.

The Contractor pays the amount of low service damages as stated in the service level table.

Example   The Scope requires the Contractor to maintain a HVAC system to a major retail development throughout the Service Period. There are specific and measurable performance criteria for the system including temperature variations and energy efficiency. The Service Period is five years.

The Scope states that the system will be maintained on a regular basis and includes a table showing how the performance of the system will be measured and acceptable levels of achievement.

The system is expected to perform to 98–100 per cent of performance criteria. If it falls within 90–98 per cent, the system will be accepted, but low service damages will be payable by the Contractor to the Client. If it falls below 90 per cent, it will not be accepted.

The service level table in Contract Data Part 1 contains the following entries:

amount per month	performance level
£500.00	96%–98%
£1,500.00	94%–96%
£2,000.00	92%–94%
£2,500.00	90%–92%

When tested, the system achieves 95% of the performance criteria.

Therefore: 360 months x £1,500.00 = £540,000 is payable by the Contractor.

OPTION X18: LIMITATION OF LIABILITY

If the Contractor causes any loss or damage to the Client or to its property, the Contractor would normally be liable for the full cost of remedial works. However, Clause X18 provides for this liability to be limited to amounts stated in the Contract Data.

In addition, the Contractor's liability to the Client for defects owing to the design of an item of Equipment may again be limited to amounts stated in the Contract Data.

Clause X18.5 can be used to place limits on the total liability the Contractor has to the Client for all matters in contract, tort or delict other than excluded matters.

Excluded matters are:

- loss or damage to Client's property;
- low service damages if Option X17 applies;
- delay damages in connection with Task Orders; and
- Contractor's share if Option C applies.

The Contractor is not liable for any matter unless it has been notified to the Contractor before the *end of liability date* which is stated in the Contract Data in terms of years after the end of the Service Period. In England, this may be six or twelve years, dependent on the type of contract, other legislations have different provisions for limitation or prescription so different limits may apply.

OPTION X19: TERMINATION BY EITHER PARTY (NOT USED WITH OPTION X11)

After the minimum period of service stated in the Contract Data a Party may terminate the Contractor's obligation to Provide the Service, for a reason not stated in the Termination Table.

If they do so, they notify the Service Manager and the other Party. The Service Manager then issues a termination certificate at the end of the notice period following the notification.

If a Party terminates after the minimum period of service, the Client may complete the service and use any Plant and Materials provide by the Contractor (P1), and the Contractor leaves the Service Areas and removes the Equipment (P4).

Payment is then made to the Contractor in accordance with A1.

- an amount due as for normal payments
- the Defined Cost for Plant and Materials which have been delivered and retained by the Client or which the Client owns and which the Contractor has to accept delivery
- other Defined Cost in reasonable expectation of completing the whole of the service (such as long-term supply contracts for consumables)
- any amounts retained by the Client

OPTION X20: KEY PERFORMANCE INDICATORS (NOT USED WITH OPTION X12)

The performance of the Contractor can be monitored and measured against Key Performance Indicators (KPIs) using Option X20.

Targets may be stated for KPIs in the Incentive Schedule.

The Contractor is required to report its performance against KPIs to the Service Manager at intervals stated in the Contract Data including the forecast final measurement. If the forecast final measurement will not achieve the target stated in the Incentive Schedule the Contractor is required to submit proposals to the Service Manager for improving performance.

The Contractor is paid the amount stated in the Incentive Schedule if the target for a KPI is improved upon or achieved. Note that there is no payment due from the Contractor if it fails to achieve a stated target.

The Client may add a new KPI and associated payment to the Incentive Schedule but may not delete or reduce a payment.

OPTION X21: WHOLE LIFE COST

Under this provision the Contractor may propose to the Service Manager that the Scope is changed in order to reduce the cost of operating and maintaining the Affected Property.

If the Service Manager is prepared to consider the change, the Contractor submits a quotation to the Service Manager which includes:

- a detailed description;
- the forecast cost reduction to the Client of the asset over its whole life;
- an analysis of the resulting risks to the Client;
- the proposed change to the Prices; and
- a revised plan showing any changes to the timing of the service.

OPTION X23: EXTENDING THE SERVICE PERIOD

The Client may, with the agreement of the Contractor, extend the Service Period, by a period up to the maximum service period stated in the Contract Data.

In order to do this, the Client notifies the Contractor and the Service Manager of the agreed extension before its notice date, again as specified in the Contract Data.

There may be certain criteria to be fulfilled for the extension. The Service Period is not extended until these criteria are met, on or before the notice date for the relevant period for extension.

OPTION X24: THE ACCOUNTING PERIODS

If this Option is used the accounting periods are stated in Contract Data Part 1.

With this Option, within thirteen weeks after the end of the accounting period (as stated), the Service Manager periodically makes an assessment of the final amount due for the service provided during the accounting period.

This assessment which provides periodic closure to costs is final unless a Party, under Option W1 or W2, refers a dispute about the assessment of the final amount due to the Senior Representatives within four weeks of the assessment being issued, refers any issues not agreed by the Senior Representatives to the Adjudicator within three weeks of the list of issues not agreed being produced, and refers to the tribunal its dissatisfaction with a decision of the Adjudicator within four weeks of the decision being made.

OPTION X29: CLIMATE CHANGE

This option allows for the Parties to include provisions for accommodating climate change requirements into the service.

Some definitions need to be explained.

(i)   The Climate Change Requirements are set out in the Scope and explain what the Contractor must do in this regard. These might include the methods, Plant and Materials or consumables that are used. It might also include recycling provisions.
(ii) The Climate Change Plan is the most up to date version of the *climate change plan* which will have been included in Contract Data Part 2 by the Contractor and subsequently updated. The Service Manager is required to accept, or not accept, each revision.
(iii) The Climate Change Partners are identified in the Climate Change Requirements and are those organisations that the Contractor will be required to work with in respect of climate change. This list will usually include other organisations working with the Client such as other contractors, tenants or users of the Affected Property and possibly external verification or regulatory agencies.
(iv) The Performance Table sets the targets that the Contractor must achieve for climate change when Providing the Service and also includes associated payments for meeting the targets. The first *performance table* is included in the Contract Data but it can be amended through various contractual routes.

X29 sets out how the Client, Contractor and others will work together to achieve the objectives. The provisions include:

a)   A requirement to collaborate
b)   The obligation to give early warnings for things that could 'adversely affect' the achievement of the Climate Change Objectives
c)   The first Climate Change Plan is either included in the Contract Data Part 2 by the Contractor or submitted for acceptance to the Service Manager within an agreed period
d)   Provisions for Defects, Contractor's proposals' and compensation events must all include the effect on the climate change arrangements
e)   Any payment due from the Contractor in respect of the Performance Table is an excluded matter if X18 has been included in the contract.

OPTION Y(UK)1: PROJECT BANK ACCOUNT

Option Y(UK)1 provides for a Project Bank Account which receives payments from the Client, which is in turn used to make payments to the Contractor and Named Suppliers.

The account holder establishes the Project Bank Account within eight weeks of the Contract Date. The account holder is either the Contractor or both Parties acting together.

Unless stated otherwise in the Contract Data, the Contractor pays any bank charges and also is entitled to any interest earned on the account.

There is also a Trust Deed between the Client, the Contractor and Named Suppliers containing the necessary provisions for administering the Project Bank Account.

The Contractor also includes in its subcontracts for Named Suppliers to become party to the Project Bank Account through a Trust Deed. The Contractor notifies the Named Suppliers of the details of the Project Bank Account and the arrangements for payment of amounts due under their contracts.

The process every month is that at each assessment date, the Contractor submits an application for payment to the Client, including the Payment Schedule, which contains details of amounts due to Named Suppliers in accordance with their contracts.

The Client makes payment to the Project Bank Account of the amount which is due to be paid to the Contractor. If the Project Bank Account has insufficient funds to make all the payments, particularly to the Named Suppliers, the Contractor is required to add funds to the account to make up the shortfall.

The Contractor then prepares the Payment Schedule, setting out the sums due to Named Suppliers. After signing the Authorisation, the Contractor submits it to the Client for signature and submission to the Project Bank.

The Contractor and Named Suppliers then receive payment from the Project Bank Account of the sums set out in the Payment Schedule after the Project Bank Account receives payment.

In the event of termination, no further payments are made into the Project Bank Account.

OPTION Y(UK)2: THE HOUSING GRANTS, CONSTRUCTION AND REGENERATION ACT 1996

This Option is only applicable to UK contracts where the Housing Grants, Construction and Regeneration Act 1996 applies and deals only with the payment aspects of the Act. Adjudication under the Act is covered by Option W2. When used with Option Y(NI)1 Y(UK)2 also complies with equivalent legislation in Northern Ireland. Similar Y clauses are published for use in Australia and the Republic of Ireland.

OPTION Y(UK)3: CONTRACTS (RIGHTS OF THIRD PARTIES) ACT 1999

This Option is only applicable to contracts in England & Wales and Northern Ireland where the Contracts (Rights of Third Parties) Act 1999 applies. Similar legislation applies in Scotland but you should take advice before using Y(UK)3 in a Scots law contract.

The principle of privity of contract is that a person who is not a party to a contract cannot enforce the terms of that contract.

The Contracts (Rights of Third Parties) Act 1999 effectively abolishes privity of contract as an absolute principle, enabling a person who is not a party to a contract, a "third party", to enforce a term of the contract if the contract expressly provides that it may. The third party must be expressly identified in the contract, either by name or as a member of a class or by a particular description, but does not need to be in existence at the date of the contract. Y(UK)3 identifies such persons as beneficiaries.

Under the Act, there is available to the third party any remedy that would have been available to it in an action for breach of contract as if it had been a party to the contract.

The parties to the contract are prohibited from making any agreement to vary the contract in such a way as to adversely affect the third party's entitlement to enforce the contract without the third party's consent.

The Act names two specific parties:

- "the promisor" means the party to the contract against whom the term is enforceable by the third party; and
- "the promisee" means the party to the contract by whom the term is enforceable against the promisor.

While the Act is not specially aimed at the construction industry, it is an industry which spends a lot of time and money in constructing collateral warranties (see Chapter 3) enabling a multitude of parties to have certain rights over specified periods of time, so it is now possible to draft building contract documents to confer benefits upon the parties who would have had the benefit of collateral warranties.

The construction industry has generally not been in favour of this act, preferring to use collateral warranties, but it must be said that the Act has several advantages in terms of time and cost in that only one document needs to be drafted, though it is fair to say that many parties prefer a warranty as specific evidence of their rights. Banks and funders in particular like the comfort of a good old fashioned document.

OPTION Z: ADDITIONAL CONDITIONS OF CONTRACT

Option Z allows conditions to be added to, amended or omitted from the existing clauses in the contract.

All changes to the original clauses should be included as Z clauses rather than amending the original clauses themselves, so in effect the clause remains in the contract, but is amended within the Z clause. It is also critical that when drafting a Z clause, it must be clearly stated what happens to the original clause, for example it is deleted. If the original clause is not deleted, then it is likely that an inconsistency will arise which will be interpreted against the party who wrote or amended the clause, i.e. the Client.

These conditions may modify or add to the original clause, to suit any risk allocation or other special requirements of the particular contract. However, changes should be kept to a minimum, consistent with the objective of using industry standard, impartially written contracts.

It must be remembered that if a Client amends a contract to allocate a risk to the Contractor which may have been intended to be held by the Client, the Contractor will price it in terms of time and money, therefore the practice of Clients amending contracts to pass risk to Contractors without considering who is best able to price, control and manage those risks can in many cases prove to be unwise and uneconomical.

It is important to spend time considering whether a Z clause is appropriate in each case, then when that decision has been made, that the clause is drafted correctly and aligned to the drafting principles of the original contract, in the case of the TSC using ordinary language, present tense, short sentences and bullet pointing.

It is not unusual to see Z clauses in a TSC contract written in a legalistic language, in the future tense and without punctuation apart from full stops.

0.11   The roles of the parties

There are a number of parties referred to in the contract:

- the Client
- the Contractor
- the Service Manager
- the Subcontractor
- the Adjudicator

The Client

Although the contract is between the Client and the Contractor, the Client is rarely mentioned in the contract.

The Client's obligations are:

- to act as stated in the contract, and in a spirit of mutual trust and co-operation;
- to allow the Contractor access to and use of each part of the Affected Property as stated on the Accepted Plan or the date for access shown on the latest accepted Task Order programme, for the Contractor to carry out the service included in the contract;
- to provides services and other things stated in the Scope; and
- to make payment to the Contractor in accordance with the Service Manager's certificates.

The Contractor

The Contractor and the Client are the Parties to the contract, the Contractor's responsibility being to Provide the Service in accordance with the Scope. The Scope describes the service, and any constraints on how the Contractor has to provide it.

There are also obligations in respect of complying with instructions, notifying early warnings, submitting plans for acceptance, and notifying and submitting quotations for compensation events which are highlighted in later Chapters.

The Service Manager

The Service Manager is appointed by the Client and manages the contract on its behalf.

The Service Manager is named in Contract Data Part 1 and may be appointed from the Client's own staff or may be an external consultant. The Service Manager can be a named individual or a company that then nominates an individual.

It is vital that the Client appoints a Service Manager who has the necessary knowledge, skills and experience to carry out the role, which includes issuing early warnings, acceptance/non-acceptance of plans, certifying payments and dealing with compensation events.

Disciplines which have carried out the Service Manager role to date include Engineers, Architects, Building Surveyors, Quantity Surveyors and Facilities Managers, in addition to Service Managers themselves.

It is also vital that the Client gives the Service Manager full authority to act for it, particularly when considering the time scales imposed by the contract. If the Service Manager has to seek approvals and consents from the Client, then it must do so, and the appropriate authority must be given, in compliance with the contractual timescales. If the acceptance process is likely to be lengthy, it is critical that both the Client and the Service Manager set up an accelerated process to comply with the contract or that the time scales in the contract are amended. The former is the vastly preferred method in order that NEC4 will work to its full effect.

Although the Service Manager manages the contract at the post-contract stage, it is usually appointed pre-contract to deal with matters such as feasibility issues, advising on procurement, cost planning, tendering and programme matters.

There should only be one Service Manager for the contract, though the Service Manager can delegate responsibilities to others after notifying the Contractor, and may subsequently cancel that delegation (Clause 14.2).

As with any other notification under the contract, if the Service Manager wishes to delegate, the notice to the Contractor must be separate to all other communications and in a form which can be read, copied and recorded, i.e. by letter or email, not by verbal communication such as a telephone call.

The delegation may be due to the Service Manager being absent for a period owing to holidays or illness, or because it wishes to appoint someone to assist with the duties.

Although the contract is not specific, the notice should identify who the Service Manager is delegating to, what their authority is, and how long the delegation will last, so the Contractor is in no doubt as to who has authority under the contract.

It is important to recognise that in delegating, the Service Manager is sharing an authority with the delegate who will then represent the Service Manager, it is not passing the responsibilities on to them. The authority is shared, but ultimately responsibility will remain with the Service Manager as the principal.

It is also important to note that one can only delegate outwards from the principal, i.e. no-one can assume authority possibly because, within an organisation, they are senior to the Service Manager.

If the Client wishes to replace the Service Manager it must first notify the Contractor of the name of the replacement before doing so.

The Adjudicator

The Adjudicator can be chosen by one of three methods:

- By the Client specifically naming the individual in Contract Data Part 1. The Contractor therefore knows at the time of tender who the Client has named and can therefore confirm if it does not agree with the Client's choice. This has advantages in that the parties always know who will be the Adjudicator if they have a dispute. The disadvantages are first that parties have commented in the past that naming the individual within the Contract Data assumes that there will be a dispute, and often the Adjudicator will require payment for having been named in the contract and being available should the parties have that dispute. Of more concern is the possibility of the named Adjudicator being unavailable at the time of need, for example through holidays or pressure of other work. The Adjudicator usually needs to reach a decision within 28 days from initial reference;

or:

- By the Client naming an Adjudicator nominating body in Contract Data Part 1. This would normally be a professional institution. This tends to be the preferred method;

or:

- By the parties agreeing who will be the Adjudicator at the time of the dispute. The disadvantage of this method is that once the parties are in dispute they probably would not readily agree with each other on anything.

The Adjudicator becomes involved only when either contracting party refers a dispute to them. As an impartial person the Adjudicator is required to give a decision on the dispute within stated time limits. If either party does not accept the decision, they may refer the dispute to the tribunal (litigation or arbitration). The Dispute Resolution Service Contract requires that payment of the Adjudicator's fees is shared by the Parties unless otherwise stated.

0.12   Some unique NEC4 TSC terms

Accepted Plan: Clause 11.2(1)

Unlike other NEC contracts, the TSC does not use the term "Programme", but a more general term "Plan". If the Service Period of the Term Service Contract is say, five years, the Contractor will not have a detailed programme for every operation.

Similarly to other NEC contracts, the Contractor must submit its Plan to the Service Manager for acceptance. If the Service Manager accepts the Plan, it is then referred to as the "Accepted Plan", the latest Accepted Plan superseding previous Accepted Plans.

Affected Property: Clause 11.2(2)

The TSC uses the "Affected Property", as the Contractor will be Providing the Service on property, over which it does not have possession or direct control, only access, such as the Client's premises. The term should be used entirely as defined, even if it sounds odd when referring to a road or railway line.

Clause 11.2(2) states that Affected Property is a property of the Client or Others, which is:

- *affected by the work of the Contractor or used by the Contractor in Providing the Service; and*
- *identified in the Contract Data, unless later changed in accordance with the contract.*

Note also the reference to Service Areas:
The Service Areas are the Affected Property and those parts of the service areas, which are:

- *necessary for Providing the Service; and*
- *used only to provide services in the contract.*

unless later changed in accordance with the contract

Equipment and Plant

The TSC definitions of Equipment and Plant are different to many other contracts.

- Equipment: Clause 11.2(7)

Equipment is defined as "items provided and used by the Contractor to Provide the Service and which the Scope does not require the Contractor to include in the Affected Property". In that sense it includes anything which is used to carry out the service, but are not incorporated into the works and do not remain on Affected Property after the Service Period has expired.

Specific examples of equipment under a Term Service Contract would include survey equipment, excavation machinery, access equipment such as scaffolding, ladders and also temporary accommodation. These items are sometimes referred to as "plant" in other contracts.

It also includes associated consumables such as fuel, lubricants and other consumables.

* Plant: Clause 11.2(11)

Plant (and Materials) are "items intended to be included in the Affected Property", specific examples being mechanical and electrical installations within buildings or overhead signage on highways.

It is critical that practitioners recognise the difference between Equipment and Plant and Materials and continue to use the correct terms properly as they are defined separately throughout the contract, and particularly within the Schedule of Cost Components and the Short Schedule of Cost Components when considering payments and compensation events.

Tasks and Task Orders

In order to consider the use of Tasks and Task Orders within the TSC, we must first consider the various related definitions:

11.2(18) A Task is work included in the service which the Service Manager instructs the Contractor to carry out and for which a Task Order programme is required.

11.2(19) Task Completion is when the Contractor has done all the work in the Task and corrected Defects which would have prevented the Client or Others from using the Affected Property or Others from doing their work.

11.2(20) Task Completion Date is the date for completion stated in the Task Order, unless later changed in accordance with the Contract.

11.2(21) A Task Order is the Service Manager's instruction to carry out a Task.

The Service Manager may issue a Task Order to the Contractor. Prior to issuing the Task Order the Service Manager instructs the Contractor to submit a quotation for the Task. The instruction includes:

* a detailed description of the work in the Task;
* the Task Starting Date and Task Completion date; and
* the amount of delay damages for late completion of the Task.

The Contractor is required to submit the quotation within three weeks of being instructed. The Service Manager replies to the quotation within two weeks of receiving it.

The assessment of the Task is in the form of a Task price list. Work covered by rates in the Price List is priced using those rates, other work is priced in the same way as a compensation event is priced.

The reply is:

* acceptance of the quotation and the issue of the Task Order;
* an instruction to submit a revised quotation;
* that the Service Manager will be making the assessment; or
* a notification that the Task will not be instructed.

The Service Manager and the Contractor may mutually agree to extend the period for the Contractor's submission and the Service Manager's reply.

If a Task Order is issued, the Task price list is inserted in the Price List, and the work involved in the Task Order is added to the Scope.

Others: Clause 11.2(9)

The TSC uses the term "Others" to define parties who are "not the Client, the Service Manager, the Adjudicator, the Contractor or any employee, Subcontractor or supplier of the Contractor". Essentially, Others are anyone outside the contract for example other contractors, statutory bodies, regulatory authorities, utilities companies, etc.

References to Others include the Contractor's obligation to co-operate with Others in obtaining and providing information, co-operating with Others and sharing the Affected Property, obtaining approval from Others where necessary, the Contractor provides access to Plant and Materials being stored for Others notified by the Service Manager (Contractor's Obligations), the order and timing of the work of Others (Time), loss or damage to Plant and Materials supplied by Others (Risk and Insurance), the Contractor has substantially hindered Others (Termination).

0.13   Communications: Clause 13

NEC4 has strict rules regarding communications under the contract:

• Each instruction, certificate, submission, proposal, record, acceptance, notification, reply and other communications must be communicated in a form which can be read, copied and recorded. The communication has effect when it is received. This may be by electronic means for example email, or via a project intranet, so the issue and receipt are simultaneous. Normally the parties agree a protocol for project communications. Where a communications system has been specified in the Scope, that is the sole way of communicating under the contract.

This requirement is particularly important in respect of instructions as the contract does not recognise verbal instructions and whereas under other forms of contract a verbal instruction can be confirmed by the Contractor and if not dissented from, normally within seven or 14 days it becomes an instruction, the NEC4 contracts do not have that provision.

• The Service Manager or the Contractor replies within the period for reply stated in Contract Data Part 1, unless the contract states otherwise. This period can be extended by mutual agreement between the communicating parties.
• If the Service Manager is required to accept or not accept, it must state the reasons for non-acceptance. Withholding acceptance for a reason stated in the contract is not a compensation event. Providing a punctual response is particularly important in respect of:

  (i) Acceptance of the Contractor's design of an item of Equipment
  (ii) Acceptance of a proposed Subcontractor
  (iii) Acceptance of the Contractor's plan and/or Task Order programme

• The Service Manager issues certificates (Payment and Termination Certificates) to the Contractor and to the Client.
• Notifications should be communicated separately from other communications (Clause 13.7), so for example, an early warning notice should not be included as part of a set of meeting minutes, or an email which includes other subjects.

0.14 Subcontracts

The TSC provides for work to be subcontracted with Clause 24 dealing with subcontracting work.
Note the TSC definition of a Subcontractor under Clause 11.2(17):

A Subcontractor is a person or organisation who has a contract with the Contractor to provide a service which is necessary to Provide the Service, except for the:

* *hire of Equipment; or*
* *supply of people paid for by the Contractor according to the time they work.*

If the Contractor proposes to subcontract work, it is responsible for Providing the Service as if it had not, and also that the contract applies as if a Subcontractor's employees and equipment were its own, thus emphasising that the Contractor is wholly liable to the Client for any Subcontractors.

Clause 24.2 is worth specific mention as the Contractor is obliged to submit the name of each proposed Subcontractor to the Service Manager for acceptance. No Subcontractor can be appointed until the Service Manager has given its acceptance.

Note again that the submission by the Contractor and the acceptance by the Service Manager must both be in a form which can be read, copied and recorded.

A reason for the Service Manager not accepting the Subcontractor is that "the appointment will not allow the Contractor to Provide the Service". This clause probably needs a word of caution to Service Managers who may give the Contractor any one of several reasons for not accepting the Subcontractor, but if the reason is not that "the appointment will not allow the Contractor to Provide the Service" then it is a compensation event under Clause 60.1(9), "the Service Manager withholds acceptance for a reason not stated in this contract".

Examples of non-acceptance of a Subcontractor which could give rise to a compensation event are "I have never heard of your proposed Subcontractor", or "I have heard that it is not doing too well on one of your other contracts", neither of which comply with the reason "the appointment will not allow the Contractor to Provide the Service".

While the contract does not state as such, clearly, if the Service Manager has concerns about a Subcontractor it should raise them with the Contractor who should be expected to provide reasonable substantiation of their competence, possibly in the form of examples of previous work and references. If, despite the Contractor providing this substantiation the Service Manager refuses to accept the Subcontractor, then it is a compensation event. An early warning might be appropriate here (see Chapters 1 and 6 later).

Service Managers are often concerned about accepting a particular Subcontractor, probably the best advice being to ask the Contractor for some substantiation and unless the substantiation proves the Service Manager's concerns, to accept the Subcontractor and remind the Contractor that under Clause 24.1 responsibility for that Subcontractor remains with the Contractor.

While the contract does not expressly require the Contractor to use an NEC contract for its Subcontractors under Clause 24.3 it is required to submit the proposed conditions of contract for the Subcontractors to the Service Manager for acceptance unless an unamended NEC contract is used, or the Service Manager has agreed that no submission is required.

As the Contractor cannot appoint the particular Subcontractor on the proposed subcontract conditions until the Service Manager has accepted them the Contractor should submit proposed conditions of contract as early as possible so that the Service Manager has a reasonable period in which to consider them.

In addition, Options C and E also require the Contractor to submit the proposed pricing information for each subcontract to the Service Manager for acceptance.

This is an important requirement as costs arising from the subcontracts will become due for payment as Defined Cost.

Nominated subcontractors

Although some other contracts provide for nominated or named subcontractors, the NEC contracts have never done so.

This is for a number of reasons:

- The Contractor should be responsible for managing all that it has contracted to do. Nomination will often split responsibilities.
- The use of Prime Cost Sums in a price list means that the Contractor does not have to price that element of the work other than for profit and attendances, the more Prime Cost Sums, the less the tenderers have to price, and thus there is less pricing competition between the tenderers.
- Contracts will often give the Contractor relief in the event of a default by the nominated subcontractor, provided that Contractor has done all that would be reasonable or practicable to manage the default. This would include the insolvency of a nominated subcontractor and a subsequent renomination.

0.15   Appendices

Service Manager duties

The Service Manager:

General

10.1	acts as stated in the contract
10.2	acts in a spirit of mutual trust and co-operation
13.3 to 13.5	replies to a communication within the period for reply
13.6	issues certificates to the Client and the Contractor
13.8	may withhold acceptance of a submission by the Contractor
14.2	may delegate any of their actions
14.3	may give an instruction which changes the Scope, a Task or the Affected Property
15.1	gives an early warning
15.2	prepares a first Early Warning Register and may instruct the Contractor to attend an early warning meeting
	may instruct other people to attend an early warning meeting
15.4	revises the Early Warning Register at each early warning meeting
16.1	consults with the Client and Contractor about a change
16.2	accepts the Contractor's proposal and issues an instruction to change the Scope, instructs the Contract to submit a quotation, informs the Contractor that the proposal is not accepted

16.3 accepts or does not accept adding to the Service Areas
17.1 notifies the Contractor as soon as aware of an ambiguity or inconsistency
17.2 notifies the Contractor as soon as aware of an illegal or impossible requirement
19.1 may issue a Task Order to the Contractor; instructs the Contractor to submit a quotation
19.3 instructs the Contractor to submit a revised quotation
19.4 extends the time allowed for submitting/replying to quotations
19.3 assesses the pricing for a Task; notifies the Contractor of the assessment

The Contractor's main responsibilities

21.1 accepts or does not accept the Contractor's design of Equipment
22.1 accepts or does not accept the Contractor's replacement people
22.2 may instruct the Contractor to remove an employee
25.3 assesses any cost incurred by the Client for Contractor not providing the services and other things
24.2 accepts or does not accept the Contractor's proposed Subcontractor
24.3 accepts or does not accept the proposed subcontract documents

Time

31.1 receives the Contractor's first plan
31.3 accepts or does not accept the Contractor's plan
32.2 accepts or does not accept the Contractor's revised plan
33.1 receives the Contractor's Task Order programme
33.3 accepts or does not accept the Contractor's Task Order programme
34.2 accepts or does not accept the Contractor's revised Task Order programme
36.1 may instruct the Contractor to stop or not to start any work

Quality Management

40.2 accepts or does not accept the Contractor's quality policy statement or quality plan
40.3 may instruct the Contractor to correct a failure to comply with the quality Plan
41.3 informs the Contractor of each test or inspection
may watch any test done by the Contractor
41.5 does tests and inspections without causing unnecessary delay
41.6 assesses the cost of repeating a test or inspection after a Defect is found
43.1 notifies the Contractor as soon as they become aware of a Defect
43.3 assesses the cost of having a Defect corrected by other people
43.4 arranges for the Client to allow the Contractor access to correct a Defect
44.1 may change the Scope to accept a Defect
44.2 accepts or does not accept the Contractor's quotation for accepting a Defect

Payment

50.1 assesses the amount due at each assessment date
50.2 considers an application for payment from the Contractor
50.4 assesses an amount due if the Contractor does not submit an application for payment
50.6 corrects any wrongly assessed amount due

51.1   certifies a payment within one week of each assessment date

53.1   makes an assessment of the final amount due; gives the Contractor details about how the amount due has been assessed

Compensation events

60.1(17) notifies the Contractor that a quotation for a proposed instruction is not accepted or that a Task will not be instructed

61.1   notifies the Contractor of a compensation event; and instructs the Contractor to submit a quotation

61.4   replies to the Contractor's notification of a compensation event
        may notify the Contractor that the Prices and Task Completion Date are not to be changed

61.5   notifies the Contractor that it did not give an early warning

62.1   may instruct the Contractor to submit alternative quotations

62.3   replies within two weeks of receipt of a Contractor's quotation

62.4   instructs the Contractor to submit a revised quotation

62.5   extends the time allowed for submitting or replying to a quotation

62.6   replies to the Contractor's notification

63.3   may agree rates or lump sums to assess the change to the Prices

63.8   assesses a compensation event as if the Contractor had given early warning

64.1   assesses a compensation event

64.2   assesses a compensation event using its own assessment of the plan for remaining work

64.3   notifies the Contractor of the assessment of a compensation event

64.4   replies to the Contractor's notification

65.1   may instruct the Contractor to submit a quotation for a proposed instruction

65.2   replies to the Contractor's quotation

65.3   may issue an instruction, notify as a compensation event, and instruct the Contractor to submit a quotation

Use of Equipment, Plant and Materials

70.1   permits the Contractor to remove Plant and Materials

Risks and insurance

84.1   receives the Contractor's insurance certificates and accepts or does not accept them

86.1   submits the Client's insurance policies and certificates to the Contractor

Termination

90.1   is notified of a reason for termination and issues termination certificate

Main Option Clauses

Option A

55.2   receives the Contractor's revision of the Price List

63.15   agrees with the Contractor to use a new People Rate

Option C

11.2(24)	decides Disallowed Cost
20.3	receives advice from the Contractor on practical implications of the Accepted Plan and subcontracting arrangements
20.4	prepares with the Contractor forecasts of total Defined Cost for the works
24.4	accepts or does not accept the Contractor's proposed subcontract Documents pricing information
54.1	assesses the Contractor's share
93.3	assesses the Contractor's share after certifying termination

Option E

11.2(24)	decides Disallowed Cost
20.3	receives advice from the Contractor on practical implications of the Accepted Plan and subcontracting arrangements
20.4	prepares with the Contractor forecasts of total Defined Cost for the works
24.4	accepts or does not accept the Contractor's proposed subcontract data pricing information

Secondary Option Clauses

W1.3(2)	may agree with the Contractor to extend the time for notifying and referring a dispute
X4:	Receives and accepts or does not accept the Ultimate Holding Company Guarantee
X10	accepts or does not accept the Information Execution Plan
X13	accepts or does not accept a Contractor's performance bond
X19	issues a termination certificate
X20	receives the Contractor's report of performance against Key Performance Indicators
X21	may consider a change
	consults with the Contractor about a quotation
X24	makes an assessment of the final amount due
X29	Accepts or does not accept the climate change plan
	Gives early warnings
	Changes Performance Table
	Notifies compensation events
	Instructs Contractor to submit compensation events quotations
	Accepts or does not accept Contractor's proposals
	Reviews Contractor's performance

1   Early warnings

1.1   Introduction

Early warnings play a key role in the management of NEC contracts, and are covered within the TSC by Clauses 15.1 to 15.4. They are also mentioned in Clauses 61.5 and 63.8, which we will discuss later in Section 1.3. There are additional references to early warnings in secondary options X10 (Information Modelling), X12 (Multiparty Collaboration), and X29 (Climate Change).

Early warnings are a key component of the overall risk management process in the TSC. The process is not about liability, but instead about the Parties collaborating to identify, mitigate or remove the effect of matters which could cause difficulty to the service provision.

Early warnings are an essential and valuable risk management tool within the NEC contracts, the TSC obliging the Service Manager and the Contractor to notify each other as soon as either becomes aware of any matter which could affect the project in terms of time, cost or quality.

Clause 15.1 requires the Contractor and the Service Manager to give an early warning by notifying the other as soon as either becomes aware of a matter which could:

- increase the total of the Prices;
- interfere with the timing of the service; or
- impair the effectiveness of the service.

The additional references in X10, X12 and X29 are for an early warning to be given of a matter that could:

X10.3   affect the creation or use of the Information Model
X12.3(3)   affect the achievement of a Partner's objectives
X29.3   adversely affect the achievement of the Climate Change Requirements

As with many NEC clauses, these ones must be interpreted carefully.

The requirement is to <u>notify</u>, and this must be done separately from other communications, in accordance with Clauses 13.1, 13.2 and 13.7. The obligation is to notify as soon as either becomes aware of a matter, and this can often be difficult for parties to demonstrate one way or the other.

Sometimes correspondence or records may show when the Contractor or Service Manager first became aware of something, but this can be a matter of subjective interpretation. Note that the contract says "*becomes aware*" not "*should have become aware*".

This sometimes causes confusion, but if we take as an example the Contractor is instructed by the Service Manager to use a particular methodology and the Contractor knows from experience

DOI: 10.1201/9781003463771-2

that the methodology would probably not be sufficient to meet the requirements of the Scope, then the Contractor should give an early warning.

The Contractor and Service Manager may also give an early warning by notifying the other of any other matter which could increase the Contractor's total cost. One could query whether a matter which could increase the Contractor's cost, but not affect the Prices, should be an early warning matter, or for that matter, whether it should be anything to do with the Service Manager, particularly if Option A has been selected, so the Contractor is not looking to recover this additional cost through the contract.

However, the words are "*the Contractor or the Service Manager may give an early warning*" so neither is obliged to do so. This provision is designed to encourage collaboration between the parties, irrespective of their contractual liability.

Note, also within Clause 15.1, there is no requirement to give an early warning for which a compensation event has previously been notified, so, as an example, if the Service Manager gives an instruction which changes the Scope, it is a compensation event, for which neither the Service Manager or the Contractor are required to give early warning.

The early warning procedure obliges people to be "proactive", dealing with risks as soon as the parties become aware of them, rather than "reactive" waiting to see what effect they have then trying to deal with them when it is often too late. Encouraging the early identification of problems by both parties, puts the emphasis on joint solution finding, rather than blame assignment and contractual entitlement.

The authors have encountered situations where Clients and Service Managers appear to be hostile to the receipt of early warnings from Contractors. Sometimes they are viewed as the first stage in a compensation event process or similar. That may be so, but not always, and in effect an early warning could prevent a compensation event occurring or at least lessen its effect.

Early warnings have a vital role to play in the TSC and their use should be encouraged, not discouraged. Clause 15.1 and the three secondary options listed earlier require the giving of early warnings; they are not optional, the Contractor and the Service Manager are obliged to issue them (read Clauses 15.1, 10.1 and 10.2 together). Clause 10.1 requires the participants to "*act as stated*" and Clause 10.2 to "*co-operate*", therefore early warnings should be issued freely and received just as freely.

Statistically speaking, one would expect an equal number of early warnings to be issued by the Service Manager and the Contractor. Typically though, the ones issued by the Contractor outnumber the ones issued by the Service Manager. There are sanctions for the failure to issue early warnings and we will discuss these later. There is no "ready reckoner" for how many early warnings should be issued but an absence of them should be viewed as more worrying than a plethora of them.

Early warnings are one of the most important and valuable aspects of the contract and it is perhaps surprising that, while a few other contracts refer to an early warning process, only the NEC contracts set out in clear detail what the parties are obliged to do, with appropriate sanctions should the parties not comply (see Clauses 11.2(6), 61.5 and 63.8).

1.2 Notifying early warnings

The contract requires (Clauses 13.1 and 13.2) that all instructions, notifications, submissions, etc are in a form that can be read, copied and recorded, so early warnings should not be a verbal communication such as a telephone conversation. If the first notification is a telephone conversation, or a comment in a meeting, it should be immediately confirmed in a suitable format to give it contractual significance.

Contract:	**EARLY WARNING NOTICE**
Contract No:	EWN No..

Section A: Notification

To: Service Manager/Contractor

Description

This matter could:

☐ Increase the total of the Prices

☐ Interfere with the timing of the service

☐ Impair the effectiveness of the service

Early warning meeting called?   **Yes/No**   **Date: ....**

Signed:...............................(Contractor/Service Manager)   Date: ....

Action by: **Date required:**

Section B: Reply

To: Contractor/Service Manager

Signed:..............................(Contractor/Service Manager)   Date: ....

Copied to:

Contractor ☐   Service Manager ☐   File ☐   Other .... ☐

Also, Clause 13.7 requires that notifications which the contract requires must be communicated separately from other communications, therefore early warnings must not be included within a long email or letter, which covers a number of issues, or embodied within the minutes of a progress meeting.

There are some key words within the obligation to notify:

- *"The Contractor and the Service Manager"* – No-one else has the authority or the obligation to give an early warning (with one minor exception in X12.3 when any of the Partners may give an early warning). The Service Manager is therefore notifying on behalf of itself, the Client and many possible others who it represents within the contract. The Contractor is notifying on behalf of itself, its Subcontractors and again many possible others who it represents under the contract.

  Service Managers are often criticised for seeing early warnings as something the Contractor has to do, and in fact as previously stated, most early warnings are actually issued by the Contractor. However, the Contractor and the Service Manager are each obliged to give early warnings each to the other, so it is critical that Service Managers play their part in the process.

  As an example, if the Service Manager becomes aware that it will be late in delivering some technical information to the Contractor, it should issue the early warning as soon as becomes aware that the information will not be delivered to the Contractor, not wait and subsequently blame the Contractor for not giving an early warning stating that it has not received the information!

- "As soon as" – means immediately. There are a number of clauses within the contract that deal with the situation where the Contractor did not give an early warning. While the party who gives the early warning must do so as soon as it becomes aware of the potential risk, the other party should respond to an instruction to attend an early warning meeting (see below) as soon as possible.

- "Could" – not must, will or shall. Clearly there is an obligation to notify even if it is only felt that something <u>may</u> affect the contract, but there is no clear evidence that it will.

It must be emphasised that early warnings are not the first step toward a compensation event as is often believed. Early warnings feature in a completely separate section of the contract and in fact the early warning provision is intended to prevent a compensation event occurring or at least to lessen its effect. It can also be used to notify a problem which is totally the risk of the notifier. It is also worth mentioning that early warnings are a notice of a future risk, not a past one. The parties are not required, nor is it of any value, to notify a risk that has already happened.

The nature and format of early warnings can have an impact on how they are received and what reaction is prompted. Contractual parties are often keen to identify the flaws of their counterparties, but the obligation here potentially requires parties to identify and record their own failures. Common sense is therefore needed in how these are identified.

The "matters" which require an early warning to be given are essentially those that affect the three determinants of success in any service provision, price, time and quality. In management terms these three are inextricably linked and will always be measured by clients of service contracts. Therefore, the contract recognises the importance of managing "matters" that may impact on these.

The first bullet in 15.1 refers to a defined term, "Prices". The definition of this term depends upon which Main Option has been selected as it is a function of the payment risk allocated between Client and Contractor. The definitions are in Clauses A11.2 (28), C11.2 (28) and E11.2 (29). Where the matter increases the Prices, an early warning must be given. Note that there is not a corresponding obligation to notify something that reduces the total of the Prices.

The remaining two bullets in Clause 15.1 are more subjective and one of the skills necessary in operating these provisions is the ability to judge how material a risk should be before it becomes the subject of an early warning. Minor risks are probably discussed, and resolved, by representatives of the Contractor and the Service Manager every day of the week but do not need to become the subject of early warnings. Many conversations are effectively short early warning meetings, but would never be formally categorised as such. Such minor matters would not be worth pursuing through Clause 15. However, where the use of an early warning would remove or mitigate the risk of an impact upon the service delivery, the early warning process should be used.

Clause 15.1 also contains an optional action for the Contractor and Service Manager, a rare item in NEC contracts. The Contractor or Service Manager may (our emphasis) give an early warning of something that could increase the Contractor's total cost. This is different from the absolute obligation to give an early warning of something that could increase the total of the Prices. It sits comfortably also with the requirements in C20.4 and E20.4 for the Contractor to provide cost forecasts to the Service Manager. Where the Contractor chooses to give an early warning in this scenario then it is likely that it is seeking assistance for a matter that at least affects itself, but possibly both Parties.

Early warnings are a proactive, not reactive, mechanism and as stated previously, it is pointless issuing an early warning of a matter that has passed.

A schedule of early warnings should be kept, identifying and numbering each early warning, with additional information about each one.

Contract: .... Contract No:			EARLY WARNING REGISTER				
Ref.	Date notified	Origin? Service Manager or Contractor	Description of event	Action still reqd by SM or Contractor?	Method of Resolution	Date Resolved	Notes
1							
2							
3							
4							
5							
6							
7							
8							
9							
10							
11							
10							

The authors have found it useful at the end of the Service Period for each Term Service Contract to review the schedule and each early warning issued during the Service Period as "lessons learned" to consider in, for example drafting Scopes for future contracts.

1.3 Remedy for failure to give early warning

One of the objectives of all NEC contracts is that the contract's provisions should act as a stimulus to good management. Typically that is as a financial incentive for doing what the contract requires or, more often, a sanction for the failure to comply. Early warnings are no exception. There are sanctions within the TSC for the failure of either the Service Manager or the Contractor to comply with these provisions.

There are express provisions covering the Contractor's failure to issue an early warning when it should have done, namely in Clauses 61.5 and 63.8. Clause 63.8 often confuses people on first reading, appearing to provide no sanction for the failure. The reverse is true. The Contractor, having failed to give an early warning, will only gain the compensation that would have accrued from the (presumably) lower cost resolution of the matter that would have followed the giving of an early warning.

Example

A Contractor is engaged under the TSC to provide building maintenance services to school buildings. It is instructed by the Service Manager to provide access to one of the buildings for independent contractors installing new IT equipment over a specific weekend. These IT contractors are directly engaged by the Client. As this is not recognised in the Scope, the Service Manager also notifies the Contractor of a compensation event.

The Contractor has also arranged for decorating subcontractors to be working in the same area of the school on the same weekend but has not told the Service Manager.

The presence of the IT staff delays the decorators and, on Monday morning, the Contractor notifies a further compensation event to the Service Manager for the increased costs of working around the IT operations.

In this scenario an early warning would have assisted the Service Manager with its planning. For example, the IT contractor could have done its work on a different weekend when the decorators were not on site. The Service Manager should therefore notify the Contractor that it did not give an early warning of this event which an experienced contractor could have given (Clause 61.5).

Then, when the compensation event is assessed it should be done as if an early warning had been given (Clause 63.8). In this scenario the disruption to the decorators could have been minimised or avoided entirely if an early warning had been given by the Contractor. Therefore the change to the Prices will be assessed on the disruption that would have been incurred had the Service Manager been able to mitigate the cost having received an early warning.

Users of the NEC contracts often query the lack of an express sanction for the failure of the Service Manager to give an early warning which it should have done. Without looking at the contract's provisions it is reasonably apparent that such a failure will, in many circumstances, lead to a cost increase or service failure for the Client and therefore the Service Manager should be incentivised to comply.

Considering Clauses 10.1, 10.2 and 15.1 together we can see that the Service Manager is obliged to give early warnings of matters that it is aware of. Failure of the Service Manager to do this could lead to a breach of contract by the Client and this will be a compensation event under Clause 60.1(15).

Early warnings often lead to an early warning meeting and Clause 15.2 allows the Service Manager and the Contractor to instruct the other's attendance at the meeting. Some commentators argue that, in certain legal jurisdictions, there is an implied obligation for the Client to attend if instructed[1]. Failure of either Party to attend would be a breach of the contract and would lead to the sanctions described above.

There are three time limits associated with the early warnings processes stated in Clause 15.2:

- The first Early Warning Register must be issued by the Service Manager within one week of the starting date.
- The first early warning meeting must be held within two weeks of the starting date.
- Subsequent early warning meetings must be held at least as frequently as stated in the Contract Data.

1.4   Dealing with risk

All commercial contracts are concerned with risk and with the allocation of risks to the Parties. NEC contracts go one step further than many other contracts and provide procedures for directly managing risk. In line with the NEC's objectives, the Parties are incentivised to follow those procedures. The early warning process is concerned with applying foresight to likely matters that could affect the delivery of the service to the end users and the Client.

Clause 80.1 defines the Client's liabilities and these are relatively standard risks taken by Clients in service contracts. Clause 81.1 then states the Contractor's liabilities.

By Clause 82, the Parties agree to indemnify one another against those liabilities. It is no coincidence that the immediately following clause relates to insurance. The Contractor's ability to indemnify the Client against the risks will most probably be underwritten by insurance.

Where a Client's liability occurs this becomes a compensation event (Clause 60.1(12)) and its impact is assessed in accordance with the compensation event provisions.

The list of Client's liabilities in Clause 80.1 is relatively standard for contracts for service provision. There is an entry in Contract Data Part 1 where the Client may list additional Client's liabilities. This entry of the Contract Data **does** amend the risk allocation of the contract, unlike the two entries which we describe later when discussing the Early Warning Register. There is not a corresponding section in Contract Data Part 2 for the Contractor to do likewise, although this could be facilitated by appropriate provisions in a tender procedure if the Client wanted to solicit such ideas.

Liabilities can be classified in a TSC contract as:

(i)   Client's liabilities which do not affect the Contractor.
(ii) Client's liabilities which do affect the Contractor in terms of Prices and are therefore included and managed as compensation events.
(iii) Contractor's liabilities which are deemed to have been included within its Fee. These are liabilities which may affect the Contractor but are not listed as compensation events.
(iv) Client's liabilities are listed under Clause 80.1, and if they occur they are compensation events under Clause 60.1(12) and which also provide for the Client to meet the Contractor's costs to others (in other words, those costs which are not Defined Cost).

Clause 80.1 also allows for additional liabilities held by the Client to be identified in the Contract Data.

Dependent on the Main Option chosen, a Contractor's liability may still be paid for by the Client, even though it will not be a compensation event.

1.5   General principles of risk management

There are a number of definitions of risk, an example being:

> *"Any uncertain event or set of circumstances that, should it or they occur, would have an effect on achieving the objectives"*
>
> *(Association for Project Management)*

The objectives in a TSC are typically to provide services safely and to the required quality, to complete within the required time scale, and within the required budget.

Risk is always inherent in any activity, no matter how simple or complex it may be. However, the degree of risk will vary depending upon the circumstances involved and the amount of risk that one is willing to assume is normally directly proportional to the amount of reward that is anticipated. The greater the anticipated reward, the greater the risk one is normally willing to assume.

Risk is an integral part of any business process . . . and an important factor in the determination of service costs and durations.

The parties to a service contract assume that the contract will not only provide them with adequate protection, but also provide a fair and workable framework within which a service can be provided to the mutual satisfaction of all concerned.

The contract itself should be clear, concise, complete, legally correct, and equitable. It should address the risks directly and attempt to eliminate as much of the ambiguity as possible.

Characteristics presenting risks

All service delivery can be regarded as having risk as an underlying factor, though there a number of differences between service provision and other areas of the economy, in that:

   (i)   every service provision is different, the only similarities sometimes being the people who manage them and the technology, materials and equipment used.

  (ii)   service contractors usually work on infrastructure designed and constructed by others.

 (iii)   construction of the property/assets may have been some time ago and there may have been inadequate service provision previously.

 (iv)   some services are delivered outside, not in a factory environment, but on a site which is subject to fairly predictable seasonal variations, but unpredictable weather exceptions.

  (v)   there is a heavy reliance upon individual specialists, particularly in ICT and electrical/mechanical systems.

As a result of the above, one can easily draw the conclusion that:

   (i)   It is impossible to specify a service wholly in advance which will not require subsequent modifications, either because of client changes or because the Contractor's plans are found in practice not to work.

  (ii)   It is impossible for a Contractor to produce a tender which will correspond exactly to the cost plus overheads and profit which will actually be incurred during the service provision.

(iii) It is impossible for a Contractor to produce a plan which can in all respects apply without revision and updating.

(iv) When people are faced with such uncertainty they may, in an effort to defend their own beliefs and organisations engage in conflict situations.

As a consequence it is clear to see that both Client and Contractor organisations are faced with risks that occur as a result of the nature of the service provision.

In addition to inherent risks within the service provision itself there are also risks that will arise from the agreement between the parties. These contractual risks are primarily expressed within the contract as intended between the parties, but they can also arise through inexperienced drafting which can lead to lack of contract clarity, ambiguities and inconsistencies.

The main principles of apportioning risk in service contracts are:

- The Client will always pay for risk, either by retaining the risk within the contract and paying for it should it occur, or the Contractor prices it within its tender.
- They should be apportioned to the party best able to manage and/or control them and to sustain the consequences if risks materialise.
- Risks outside the Contractor's control and/or foreseeability should remain with the Client, but commercial reality usually prevents this from happening fully.
- The party carrying the risk should be motivated to manage the risk, possibly through a risk/ reward process.

The major types of risks that can arise in service contracts:

Technical

- Ability of existing design to deliver service of sufficient standard and quality.
- Access to intellectual property necessary to manage infrastructure.

Logistical

- Availability of resources
- Ability to avoid business interruption

Resources

- Insufficient resources and expertise.
- Weather and seasonal problems
- Industrial relations problems

Timing

- Inability to give access to the contractor

Financial/Contractual

- Inflation
- Delay in payment

- Unavailability of funds
- Cash flow problems
- Ambiguities and/or inconsistencies in contract documents.
- Work carried out by directly employed parties, e.g. statutory undertakers
- Losses as a result of default of contractors

Safety/Reliability

- Delays as a result of having to satisfy requirements

Quality Assurance/Standards

- Quality issues

Political

- Changes in Government/new legislation
- Customs and import restrictions

Inflation

- Differential inflation owing to market situation at time of tendering
- Abnormal inflation

Errors in tender documents
Variations

- Effect on timing of service delivery and price of late variations

Force majeure

- Fundamental risk, which cannot be priced

The Risk Management Process

The allocation of Client's liabilities and Contractor's liabilities reflects the Parties' allocation at the Contract Date. As events arise during the delivery of the services, the contract will allocate responsibility and will sometimes define a process for the Parties to follow.

Early warnings are one of those processes, focused on the resolution of the problem rather than the blame or allocation of liability. The authors have seen repeated successful contracts from parties who understand the role and purposes of the early warnings, Early Warning Register and early warning meetings. These three tools dovetail with one another and act as a benefit to parties who embrace them.

Risk management is a structured approach to identifying, assessing and managing risk. It can be defined as:

"The systematic application of management policies, procedures and practices to the tasks of identifying, analysing, assessing, treating and monitoring risks".

The risk management process can help a party analyse the degree of uncertainty in achieving the contract objectives, understand the consequences of particular events occurring, and develop management actions to control the event or minimise its consequences.

There are six primary activities in the process:

1. Risk Identification

Risks should be identified as early as possible and continuously monitored to highlight new risks and the passing of old risks. Risk identification should initially be carried out by analysing all available documentation and knowledge as a brainstorming session involving a number of individuals, ideally chaired by an independent person external to the project. This will ensure that:

2. Risk Analysis

Once the risks have been identified they should be classified in terms of:

*   likelihood of occurring
*   impact on the contract objectives should they occur.

Each risk can then be given a ranking, possibly:

*   Extremely unlikely
*   Unlikely
*   Moderate
*   Likely
*   Almost certain

Impact can also be given a ranking, possibly:

*   Insignificant
*   Moderate
*   Major
*   Severe
*   Catastrophic

3. Risk Planning

This is the process of developing responses to address, manage or control risks. The responses will consider:

*   **Avoiding the risk:** fundamental changes to the way the service is delivered or, in extremes, the Client bringing the service back 'in-house'?
*   **Reducing the probability**: doing something a different way or educating the people that may be subject to the risk, so the risk is less likely to occur?
*   **Reducing the impact:** including resources, funds and/or float in programmes to cushion the effect of the risk?

- **Transferring the risk:** passing the risk to another (willing) party or an insurance company who is best able to manage or bear the risk?
- **Accepting the risk:** with the full knowledge of what the risks and potential outcomes are.
- **Ignoring the risk:** with the full knowledge of what the risks and potential outcomes are.

4. Risk Tracking

Risks should be tracked to monitor the status of each risk as the contract progresses.

5. Risk Controlling

An activity that utilises the status and tracking information about a risk or risk mitigation effort, a risk may be closed or watched, a contingency plan may be involved.

6. Risk Communicating

An action to communicate and document the risk at all times. This will normally be done in the form of a risk register.

When considering risk it is vital to consider:

- Which party can best foresee the risk
- Which party can best bear the risk
- Which party can best control the risk
- Which party most befits or suffers if the risk materialises

1.6   The use of an Early Warning Register under the TSC

The NEC3 contracts provided for the use of a Risk Register, the NEC4 contracts referring to it as an "Early Warning Register".

The Contract Data does contain matching sections in Parts One <u>and</u> Two for the Parties to add matters that will be included in the Early Warning Register. That is to say that both Parties have the ability to list matters that they wish to form part of the risk management processes in the contract. These, as we will see, <u>**do not**</u> change the risk allocation between the Parties. These entries are to assist the risk management processes of the Parties by listing those risks which collaborative behaviour will help with.

The Early Warning Register is a simple document, but a vital one in the process of risk management. Its role and contents are sometimes misunderstood by Parties particularly where people confuse its use with other components of company management systems. Its purpose is to assist the Parties and the Service Manager with managing risk in the service provision and not to allocate responsibility or blame.

It is helpful to start with its definition from Clause 11.2(6):

The Early Warning Register is a register of matters which are:

- *listed in the Contract Data for inclusion;*
- *and notified by the Service Manager or the Contractor as early warning matters.*

It includes a description of the matter and the way in which the effects of the matter are to be avoided or reduced.

So, read the definition carefully, the Early Warning Register contains information about these matters:

- Those listed in the Contract Data
  - by the Client in Contract Data Part 1
  - by the Contractor in Contract Data Part 2
- Those notified as an early warning matter
  - by the Service Manager
  - by the Contractor

The information contained in the Early Warning Register is:

- a description of the risk
- the actions to be taken to avoid or reduce the risk

The Early Warning Register need not contain anything else. It is a document produced after the Contract Date and therefore it does not form part of the contract.

For practical reasons though it should not contain any statements that purport to allocate risk to one party or the other, that will just prove confusing in the long run.

It is critical that the parties fully understand that the purpose of an Early Warning Register is to list all the identified risks and the results of their analysis and evaluation. It can then be used to track, review and monitor risks as they arise to enable the successful completion of the project. The Early Warning Register does not allocate risk; that is done by the contract.

The contract does not prescribe the format or layout of the Early Warning Register, other than to list the risks in the contract and those that come to light at a later date and are notified as early warnings following which, if there is an early warning meeting, the Service Manager revises the Early Warning Register to record the outcome of the meeting.

Note that by the definition within Clause 11.2(6), matters which were not originally included in the Contract Data or subsequently notified as early warnings should not be included in the Early Warning Register. Contract Data Part 2 allows the Contractor to identify matters which will be included in the Early Warning Register. The Early Warning Register does not allocate or change the risks in the contract, it records them and assists the parties in managing them. In that sense it is a valuable addition to the workings of the contract.

Figure 1.3 shows an extract from a typical register which would comply with the contract in that it includes the basic requirement for a description of the matter and a description of the actions which are to be taken to avoid or reduce the effects of the matter.

There is no stated list of components of an Early Warning Register, but column headings should typically be titled:

(i) Description of matter

A clear description of the nature of the matter, if necessary referring to other documents such as a report of a test or survey, etc.

(ii) Implications

What would happen if the matter were to occur?

EARLY WARNING REGISTER

Contract: Property Inspections and Repairs - Apex Housing Estate

Contract No: 289

Contractor: Buildit Ltd
Service Manager: John Smith

Description of Risk	Implications	Likelihood of Occurrence (1–5) (Least - Most)	Potential Impact (1–5) (Low - High)	Risk Score	Risk Owner	Mitigation Strategy By whom? By when?	Allowance in the total of the Prices	Programme Allowance	Client cost allowance	Risk Status	Last Updated
Possibility and extent of defects in domestic heating systems	Additional resources and costs in rectifying	3	4	12	Client	Contractor to inspect and report to Service Manager on weekly basis, to assess need for further Task Orders	Allowance in Client budget	No allowance	£100k	Rising. Early indications show heating systems are more defective than originally anticipated	9/15/2017

Copied to:

Contractor

Service Manager   File   Other . . . .

(iii) Likelihood of occurrence

This provides an assessment of how likely the risk is to occur. The example shows a forecast on a 1 (least likely) to 5 (most likely) basis, though it may be assessed as percentages, colour coding, or simply "Low" (less than 30% likelihood), "Medium" (31–70% likelihood), "High" (more than 70% likelihood).

(iv) Potential impact

This assesses the impact that the occurrence of this matter would have on the project in terms of time and/or cost. The example shows the assessment on a 1 (low) to 5 (high) basis.

(v) Risk score

Risk = Likelihood of something occurring x Impact should it occur.
    By evaluating risk on that basis each can be evaluated and categorised into an order of importance. This formula can only be applied to economic loss, and different standards would need to be adopted when considering risk of death or serious bodily injury.

(vi) Mitigation strategy

This registers what actions are proposed which could be taken to prevent, reduce, or transfer the risk of the matter occurring. Also, who is responsible for this action, and when should the action take place?

(vii) Risk status

This identifies whether the risk is current, and also whether it is increasing, decreasing, or has not changed since it was last reviewed?

(viii) Last updated

When was an additional matter identified? When was the Early Warning Register last updated?
    It is possible, but not the intention of NEC's drafters, for the Contractor to qualify its tender by including additional matters within Contract Data Part 2 which it sees, or would like to transfer as Client's liabilities. That is not the intent of the TSC as the only place to amend the risk profile is in Contract Data Part 1. However, the wording of the final bullet of Clause 80.1 leaves open the possibility of the Contractor inserting such an entry into Contract Data Part 2. Clients should check for such entries during tender appraisal.
    Risk management is essential to the success of any service delivery and normally follows five steps:

    Step 1: Identify the risks
    Step 2: Decide who could be harmed and how
    Step 3: Evaluate the risks and decide on precautions
    Step 4: Record the findings and implement them
    Step 5: Review the assessment and update if necessary

In order to compile the Early Warning Register the matters must first be listed, then they are quantified in terms of their likelihood of occurrence and their potential impact upon the project.

1.7 Early warning meetings within the TSC

The Early Warning Register becomes more important once early warnings lead to early warning meetings, see Clauses 15.2, 15.3 and 15.4.

Under Clause 15.2, either the Service Manager or the Contractor may instruct the other to attend an early warning meeting. Subcontractors may attend if attendance would assist with resolving the matter. Other participants may attend with the consent of the Contractor and Service Manager.

An early warning meeting may only have the Service Manager and the Contractor present, though each may instruct other people to attend if the other agrees, so in reality a number of people normally attend the meeting. The "rule of meetings" will often apply in that the productivity of the meeting is often inversely proportional to the number of people who attend it! But this should not mask the importance of getting the right people to attend. Service delivery, particularly in buildings, relies on multiple specialists and these should be encouraged to participate wherever necessary. The use of online meetings is now routine, reducing the need for people to incur travel time to attend.

Clearly the purpose of the early warning meeting is to actively consider ways to avoid or reduce the effect of the matter which has been notified. In some cases, the matter can be fully resolved, but in others, as the matter may not yet have occurred it may simply have to be "parked" and recorded as such in the Early Warning Register.

Reading the contract we can see that the Service Manager has a duty to enter the early warning matter into the Early Warning Register once it has been notified. Typically this is part of the communications system referred to in Clause 13.2 so should be immediately available to others working on the contract.

There follows another rare example of an optional action in an NEC contract, when either the Service Manager or the Contractor <u>may</u> (our emphasis) instruct the other to attend an early warning meeting. Hopefully in the real world such meetings are arranged by co-operation and invitation. This clause is clearly intended to promote ownership of the service delivery by the Contractor and the Service Manager, and any issues that could affect it, together with their resolution.

Early warning meetings can happen in several ways;

- As a standing agenda item in regular (say weekly, or monthly) meetings
- As a routine meeting (say once a month, as set out in the Contract Data)
- Ad-hoc, when the matter is sufficiently urgent

Meetings can provide a distraction to the "day job" so complying with Clause 15.2 should not place unnecessary demands on the participants. The use of telephone or online conferencing works well, particularly with multiple participants in different locations.

Clause 15.3 describes the agenda for an early warning meeting and it is a relatively predictable risk management process which leads to the Service Manager and Contractor agreeing actions to remove or mitigate a risk.

The contract requires those attending to co-operate in:

- Making and considering proposals
- Seeking solutions

- Deciding on actions
- Deciding which risks have been avoided or passed and can therefore be removed from the Early Warning Register.

This final bullet point is the conclusion of the process for individual matters. Clause 15.4 goes on to say that the Service Manager revises the Early Warning Register after the risk reduction meeting. While Clause 15.3 mentions removing risks from the Early Warning Register we recommend retaining them, perhaps in an appendix or struck through, for future consultation.

So, what do those attending do in an early warning meeting? The purpose is to reduce or remove risks that may affect cost, time or quality of the service provision.

Example

A Contractor maintaining airside infrastructure at an airport has to undertake electrical testing to aircraft grounding lights. This requirement is included as a part of its planned preventative maintenance obligations and the testing is part of statutory requirements for operating the airport.

The Contractor gives an early warning to the Service Manager that heavy rain is forecast for the next week when these tests are due to be undertaken which may make progress slower than anticipated or, in the worst case, prevent the work from happening at all. The Service Manager entered this early warning into the Early Warning Register.

The early warning meeting was held by web conference and included the Service Manager, the Contractor, the Airfield Operations Manager and the specialist electrical subcontractor that was appointed to do the tests. The meeting identified which lights were (a) approaching their expiry date for the current certificate and (b) operationally necessary for the next week. Having done this analysis, the workload for the week was prioritised ensuring that the most operationally necessary work was completed and that abortive costs were reduced.

The Early Warning Register was revised by the Service Manager after the meeting to record the decisions made which provided a reference for similar situations in the future.

It is the Service Manager's responsibility to revise the Early Warning Register to record the outcome of the meeting, and to issue the revised Early Warning Register to the Contractor, although this process is often semi-automated with online communications systems.

1.8   Contractor's proposals

Clause 16 is a new provision, where the Contractor may propose to the Service Manager that the Scope is changed to reduce the amount the Client pays to the Contractor. This could also be considered as part of a risk management process.

An example could be that a material specified within the Scope could be substituted with another, less expensive material that is equally effective, or a specified methodology could be substituted with a more rational, and again less expensive methodology, for example the use of mobile access platforms rather than a full static scaffold.

Within 4 weeks of making the proposal, the Service Manager may:

- accept the proposal and issue an instruction changing the Scope;
- inform the Contractor that the Client is considering the proposal and instruct the Contractor to submit a quotation; or
- inform the Contractor that the proposal is not accepted.

The Contractor benefits from such a proposal:

- In Main Option A contracts, the Contractor receives a share of the saving through the value engineering percentage (Clause 63.12).
- In Main Option C contracts, the Defined Cost of the service will reduce but the Prices are unchanged, so the Contractor's share will improve (Clause 63.13).

The Contractor may also submit a proposal for adding to the Service Areas to the Service Manager for acceptance.

The implications of adding to the Service Areas depend on the Main Option chosen:

Option A

The Short Schedule of Cost Components refers to the cost of resources used within the Service Areas, so these resources if also used within the extended service areas would be included as cost rather than within the fee percentage when assessing compensation events.

Options C and E

The Schedule of Cost Components refers to the cost of resources used within the Working Areas, so these resources if also used within the extended working areas would be included as cost rather than within the fee percentage for payments, as well as when assessing compensation events.

A reason for the Service Manager not accepting is that the proposed area is:

- not necessary for Providing the Service;
- used for services not in the contract

2 The Contractor's main responsibilities

2.1 Introduction

The Contractor's main responsibilities are covered within the TSC by Clauses 20.1 to 27.2 with further references within the Main Option Clauses.

The Contractor is obliged to Provide the Service in accordance with the Scope.

The Scope will include a detailed description of the services to be provided, possibly including drawings, specifications and schedules, and any constraints in providing the services, for example working hours, access issues and permissions, etc. It should also include how success, or failure, in Providing the Service is to be tested and measured.

Any information on the infrastructure receiving the service (referred to as the "Affected Property"), and surroundings, if relevant, must also be included within the Scope. This could include information on utilities information (electricity, gas, water and telecommunications) regarding existing installations and also environmental issues such as protected species, etc.

In Providing the Service, the Contractor is obliged to minimise the interference caused to the Affected Property and the activities taking place in it.

Also what is (or is not) provided by the Contractor and the Client in terms of services, facilities and other things should be clearly defined with the Scope.

2.2 Design of Equipment

The TSC has express provisions for the Contractor to carry out design of Equipment, which includes temporary work, for example erecting scaffolding for operations at height.

However, it should be noted that it is not uncommon to include design responsibilities within a term service contract, so Z clauses may need to be considered when drafting the TSC in order to include for that provision.

If the Contractor has designed an item of Equipment, for example a temporary access way, or specialist scaffold, the Service Manager may instruct it to submit particulars for acceptance (Clause 21.1), so the submission is not obligatory.

The Service Manager cannot accept the design if it does not comply with:

- the Scope;
- the Accepted Plan; or
- the applicable law.

DOI: 10.1201/9781003463771-3

2.3   People

Key people

The Contractor is required to name each "key person" within Contract Data Part 2, including their name, job, responsibilities, qualifications and experience. The requirement to identify such people is usually set out in tender documents.

In Providing the Service the Contractor is required to employ the same people or a replacement which has been accepted by the Service Manager, including the requirement to submit details of the replacement's name, job, responsibilities, qualifications and experience. The Service Manager may refuse to accept the replacement, as their qualifications and experience are not as good as the person they are replacing.

Removing a person

Under Clause 22.2, the Service Manager may instruct the Contractor to remove a person provided by the Contractor. The Service Manager must state the reasons for giving such an instruction.

Many practitioners have queried what sort of reasons a Service Manager could give for removing a person, typical reasons usually being:

- the person has infringed a health and safety requirement, possibly creating a danger to themself and to others.
- the person is not competent to carry out its duties.
- the person has behaved inappropriately towards the Client, the Service Manager, or another party.

Clearly the Service Manager cannot give a reason that is illegal under the law, for example instructing the Contractor to remove a person purely on the basis of their race or gender.

An interesting example used once, was that the Service Manager instructed the Contractor to remove a person as they did not like him, which, unless it is considered as a breach of the obligation under Clause 10.2 to act in a spirit of mutual trust and co-operation can only be considered as a valid reason!

2.4   Working with the Client and Others

The TSC uses the term "Others" to define parties who are "not the Client, the Contractor, the Adjudicator or any employee, Subcontractor or supplier of the Contractor", which could include other contractors, utilities companies, third party regulators, etc.

Clause 23.1 requires the Contractor to co-operate with Others in obtaining and providing information which they need in connection with the services. Under Clause 25.1, the Contractor is also required to obtain approval from Others where necessary.

The Contractor may also have to share the Affected Property with Others as stated in the Scope. The Scope may also require the Contractor to co-ordinate the work with others including holding or attending meetings with them, and notifying the Client before the meetings, in case the Client wishes to attend them.

2.5   The Client's obligations

Clause 23.2 covers the Client's obligations to provide services and other things which the contract requires it to provide in accordance with the Scope.

The use of the terms "thing" and "things" is a little curious, and usually applies to something that cannot be adequately described! However, it applies to facilities, services etc. which the Client provides, and should be fully described in the Scope and also specifically referred to in Contract Data Part 1 which identifies what access is to be given, and when.

It is critical that the Parties understand what is to be provided by the Client, and also it is critical that, if there is a charge for the availability and use of those things it should be clearly detailed within the contract, including how the charge will be measured and when the Contractor has to pay.

The Service Manager may give an instruction to the Contractor which changes the Scope, a Task or the Affected Property (Clause 14.3).

Contractors (and Consultants) are normally required to adhere to Codes of Conduct which include a duty to behave ethically, and are normally set by the professional institution or trade bodies of which they are members.

Examples of Codes of Conduct normally include obligations to:

(i)   discharge professional duties with integrity and in a fair and unbiased matter, to uphold and enhance the standing and reputation of the relevant institution.
(ii) only undertake work that they are professionally and technically competent to do and in carrying out that work to adhere to good practice and to demonstrate a level of competence consistent with a professional of their standing. This may also be coupled with the requirement to maintain insurances and indemnify their Clients against risks.
(iii) have full regard for the public interest, particularly in relation to matters of health and safety, but also to others' business activities, and to the well-being of future generations.
(iv) show due regard for the environment and for the sustainable management of natural resources.
(v)   develop their professional knowledge, skills and competence on a continuing basis and give all reasonable assistance to further the education, training and continuing professional development of others.
(vi) not improperly offer or accept gifts or favours which would be interpreted as exerting an influence to obtain preferential treatment.

Examples of a Service Manager giving an instruction which could require the Contractor to act outside its professional code of conduct would be instructing something which was outside the Contractor's area of expertise, which could cause harm to the Contractor or to others, or to do something which would infringe another's copyright, or put it in a position where there would be a conflict of interest.

2.6   Scope provided by the Client v Scope provided by the Contractor

The TSC is unlike many other commercial contracts in that it does not provide a priority or hierarchy of documents, but if one "reads into" the various clauses you may determine which takes precedence in the event of an inconsistency.

A question to then be considered is, if the Contractor has provided Scope for its plan, and there is an inconsistency between the Scope provided by the Client and the Scope provided by the Contractor for its plan, which takes precedence?

The answer can be found within Clauses 11.2(5) and 60.1(1):

A Defect is defined in Clause 11.2(5) as:

- a part of the service which is not in accordance with the Scope; or
- a part of the service which is not in accordance with the applicable law or the Accepted Plan.

Clause 60.1(1): The following events are compensation events [includes]:

The Service Manager gives an instruction changing the Scope <u>except</u>:

- a change made in order to accept a Defect; or
- a change to the Scope provided by the Contractor for its plan which is made:
- at the Contractor's request; or
- in order to comply with the Scope provided by the Client.

2.7   Fitness for purpose v reasonable skill and care

Where Task Orders are issued under a TSC to design and/or construct further infrastructure to add to the Affected Property, we need to consider the liability taken by the Contractor.

In order to consider the responsibilities of contracting parties, we should consider the principles of "fitness for purpose" and "reasonable skill and care".

While these terms are normally considered in terms of design responsibilities, they can also include other services, and workmanship, both of which could form part of a TSC.

If one considers a building project as an example, "fitness for purpose" means producing that building, part of a building, or a product fit for its intended purpose.

This is an absolute duty independent of negligence. In the absence of any express terms to the contrary, a Contractor or Subcontractor who has design responsibility, if the contract provides for it, will be required to design "fit for purpose".

"Reasonable skill and care" means performing to the level of an ordinary exercising a particular skill. The criteria is that it is not possible or reasonable to judge someone's professional ability purely by results; for example, a lawyer whose Client is found guilty is not necessarily a poor lawyer, and a surgeon whose patient dies is not necessarily a poor surgeon. The question to ask is "did that professional apply the correct level of skill and care in carrying out their duties?"

If one then applies the same criteria to a Contractor designer, this is normally achieved by the designer following accepted practice and complying with statutory requirements, codes of practice, etc, but if a new construction technique is involved the duty can be discharged by taking advice from Contractors and warning the Client of any risks involved. This is the "state-of-the-art" defence, meaning that a designer is only expected to design in conformity with the accepted standards of the time.

In the absence of any express terms to the contrary, a Contractor will normally be required to design and carry out other services using "reasonable skill and care".

Design and build contracts will often impose a higher standard than reasonable skill and care on Contractors, i.e. fitness for purpose. This obligation resembles a seller's duty to supply goods which are fit for their intended purpose. The dilemma for design and build Contractors is then, that they are bound to a "fitness for purpose" obligation, whereas the consultant that did that design on its behalf is usually limited to "reasonable and care", a dilemma that continues to haunt design and build Contractors where a design defect arises!

It is worth noting that the "default" position within the NEC contracts is that any design carried out by the Contractor must in all respects be fit for purpose. As stated above, this means producing a building, part of a building, or a product fit for its intended purpose. This is an absolute duty independent of negligence. If parties to a TSC wish to amend this then similar provisions from the NEC4 Engineering and Construction Contract clause X15 could be incorporated via a Z clause.

In the absence of any express terms to the contrary, a construction contract which includes design requires the Contractor to design 'fit for purpose'.

In a TSC contract, these issues would become relevant if the Contractor was given a Task Order requiring it to design and construct additional infrastructure. For example, a highways maintenance contractor instructed to re-configure an existing junction or a schools maintenance contractor being required to install new power supplies as part of IT upgrades.

2.8   Collateral warranties and undertakings to Others

Collateral warranties may need to be considered within a TSC to form a relationship between the Client and, maybe a specialist Subcontractor?

A collateral warranty is an agreement associated with another contract, collateral meaning "additional but subordinate" or "running side by side", the collateral warranty being entered into by a party to the primary contract, for example a Contractor or a Subcontractor, and a third party who is not a party to the primary contract but who has an interest in the service provided.

Collateral warranties came into popular use following some notable court cases in which it was judged that it was not possible to recover damages for negligence in relation to defects in contracts as it was held to be economic loss, which is not recoverable in tort.

Consequently, such claims had to be brought as claims for breach of contract, where economic loss is recoverable. It therefore became necessary to establish a contract between parties involved in a project or service, such as designers or subcontractors and parties such as funders, present and future purchasers, who are not parties to the primary contracts but who have an interest in the project, hence the "collateral warranty".

The NEC4 TSC provides for these arrangements in secondary option X8. NEC contracts do not use the words "collateral warranty", instead referring more descriptively to "Undertakings to the Client or Others." When utilised, X8 can require the Contractor to enter into additional agreements with third parties and for its Subcontractors to do likewise with third parties and/or the Client.

Such undertakings are different to traditional contracts, in that the two parties to the warranty do not have any direct commercial relationship with each other. They are useful for creating ties between third parties, but the main disadvantage of collateral warranties is the expense of providing them, particularly where there are many interested third parties such as with housing and commercial properties. When one also considers that parties can change their names or business status during the life of a warranty then even a medium sized term service contract can involve 30 or more warranties.

Different parties will also have different requirements for collateral warranties, therefore the precise terms of the collateral warranty agreement need to be considered in each case carefully so that they are drafted to provide the required protection. In the TSC the wording of such agreements should be included in the Scope.

Collateral warranties are executed as a "deed" also referred to as "under seal", which differs from a simple contract in two respects:

* First, in English Law the time after a breach of contract has occurred within which one party can sue another is 12 years, whereas the time for a simple contract opposed to 6 years for an agreement that is executed "under hand".

- Second, whereas under English law consideration is needed for a contract to be effective, this is not the case with a deed. For a company to execute a deed effectively, the document must be signed by two directors or a director and the company secretary, unless the Articles of Association contain some other requirement.

The basic ingredients of a collateral warranty are usually:

(i) Definitions

There will be a brief description of the work and the collateral warranty. The "contract" is the contract to which the collateral warranty agreement is collateral and the date should be specified. Since the obligations of the Contractor to the beneficiary parallel the obligations of the contractor to the client under the contract for the core service, the beneficiary should always ask to see a copy of the contract. If, for example, there are restrictions on the Contractor's liability in the principal contract, that might affect the rights of the beneficiary under the collateral warranty.

A copy of the principal contract may be attached to the collateral warranty agreement.

(ii) The warranty

The Contractor gives warranties to the beneficiary that it has exercised and will continue to exercise reasonable skill and care in its obligations to the Client under the principal contract including compliance with any design obligations, performance requirements and correction of defects.

The warranty should contain a statement that the warrantor shall have no greater liability to the beneficiary than they would have under the primary contract.

There should be a statement that the warrantor shall have no liability under warranty in any proceedings commenced more than 12 years after the date of completion of the works.

(iii) Naming of the parties

The named parties can be, for example, the Contractor and the end user, a Subcontractor and the Client, a Contractor and the Client.

The party that gives the warranty is known as the warrantor, the other party is known as the beneficiary.

(iv) Prohibited materials

The Contractor may be required to give a <u>warranty</u> that it will not specify or use materials which are known to be prohibited or deleterious to health and safety.

In addition, the use of deleterious materials may contravene relevant legislation or regulations as well as being expressly prohibited in the contract between the developer and the contractor.

(v) Copyright

Normally, the beneficiary under a collateral warranty agreement will have a right to use designs and other documents prepared by the Contractor but only in connection with the Affected Property for which those design documents are prepared, copyright to these documents remaining with the Contractor.

(vi) Insurances

There is normally a requirement for the warrantor to maintain insurances including professional indemnity insurance where applicable for a stated minimum level of cover and for a period of 12 years after completion of the works.

(vii) Assignment

There is normally a provision allowing the beneficiary up to two assignments of the warranty.

(viii) Step in rights

Step in rights can allow a funder to take over the Client's role and become the Client under the primary contract. This is particularly important where the Client is unable to pay the Contractor, the Contractor having to notify the beneficiary before it terminates the contract, this allowing the beneficiary to "step in" and take the Client's place, with an obligation to pay the Contractor any money that is outstanding. The beneficiary does not have any obligation to step in, only the right to do so. As the effect of this clause is to vary the principal contract it is important that, when the collateral warranty contains such a clause, the Client should be a party to that document.

(ix) Law and jurisdiction

It is important to specify the law of the contract and the warranty and the jurisdiction of the courts will have jurisdiction.

(x) Dealing with planning permission

The responsibility for obtaining planning permission can lie with the Client or can be assigned to the Contractor. This might be important if a Task Order required the design and construction of additional infrastructure.

It is more usual the Client to gain outline planning permission and the Contractor to be responsible for gaining full and detailed planning consent, but the Parties should remain aware of clause 25.1.

2.9   Assignment

Under Clause 26, either Party must notify the other Party if they intend to transfer the benefit of the contract or any rights under it.

In addition, if the Client cannot transfer a benefit or a right is the party receiving it does not intend to act in a spirit of mutual trust and co-operation.

See the earlier comments in the Introduction to this volume about the interpretation, and difficulty of enforcing, mutual trust and co-operation provisions in a contract. In this case, how would the Client prove (or someone else disprove) that the other party intends to act in a spirit of mutual trust and co-operation? Also, one can always "intend" to do something, but later fail to comply".

3 Time

3.1 Introduction

Time is covered within the TSC by Clauses 30.1 to 36.1.

A successful service is considered by clients and contractors to be one which has been completed on <u>time</u>, to the required <u>quality</u> and also within the financial <u>budget</u>. . . . .and by contractors and subcontractors as one which has been completed on <u>time</u>, to the required quality and making a <u>profit</u>. Clearly, apart from their different perspectives on financial outcome, the objectives of both contracting parties are well matched.

When considering completion on time, various surveys carried out in recent years have shown that only about 33 per cent of contracts are completed within the allotted time period, one fifth overrun by 40 per cent and a significant number overrun by more than 80 per cent!

There are many factors that can cause delays:

  (i) Inability to give timely access
 (ii) Non-availability of resources, by all parties
(iii) Poor performance of parties
 (iv) Client changes, compounded by the number or timing of instructions
  (v) Weather and other seasonal problems
 (vi) Insolvency of a party
(vii) Unrealistic time periods set by clients

Clearly matters need to be improved by the parties and between the parties, but many of these factors can be avoided or at least reduced by the parties engaging in proper and effective planning and programming of the service aligned to effective communication between the parties.

Planning and programming

The terms "planning" and "programming" are synonymous. However, they are a little different.

  (i) What is planning?

Planning may be defined as "*the deliberate consideration of all the circumstances in order to evolve the best method of achieving a stated objective*". This involves defining the service, setting the overall duration, and planned resources, and identifying risks.

Planning is a vital element in any business process, without which it is impossible to envisage the successful conclusion of any activity.

DOI: 10.1201/9781003463771-4

Planning considers. . . . .

- What to do
- When to do it
- How to do it
- What resources are required to do it
- Who is going to do it
- How long it will take to do it

The motto is. . . . .

PLAN YOUR WORK. . . . . AND WORK TO YOUR PLAN!

Planning may be considered on a long or short term basis, long term planning including the development of a master plan for the whole service, while short term planning is a much more detailed operation, which considers only the next 3−4 weeks of operations, or a particular element of the service. It is essential that each short term plan should overlap the other by 1−2 weeks to allow for past and future activities to be considered.

The planning process considers each operation in turn, analysing the order and sequence of each operation, its dependency on previous operations, and the contractor's and client's resources to carry it out and the availability of those resources.

It is easier to track progress of operations using short term, rather than long term planning.

It is considered that there are three planning stages:

(a) Pre-tender planning

This covers the planning considerations during the preparation of a cost estimate and its conversion into a submitted tender price.

The process enables the contractor to:

- establish realistic periods on which the tender may be based. The client may set the starting date and the completion date for individual activities.
- consider the most effective economical methods, using the most effective resources.
- assist the build-up and pricing of contract mobilisation charges and equipment expenditure.

This is usually carried out in broad outline format as the contractor may not be successful in its tender.

Typical information which may be gathered and used at the tender planning stage includes:

(i) A brief description of the service and its location, details of the client, and the various consultants, key companies, third parties and individuals.
(ii) Details of any planning issues or restrictions.
(ii) Access to site for delivery and collection of equipment, plant and materials and also removal of waste.
(iv) Details of existing services and the location of new and existing connection points. Details of the relevant service providers.
(v) Availability of people, equipment, plant and materials.
(vi) Details of possible subcontractors.
(viii) Typical weather conditions.

This information is then collected together with information given to the contractor at tender stage to develop a pre-tender plan. Time-based elements can then be transferred to a pre-tender programme, usually in a bar chart format which together with the method statements then aids the estimator in preparing the tender and assessing the need for mobilisation charges and major multi-use equipment.

Clients sometimes instruct contractors to submit plans with their tenders, though it must be appreciated that as the contractor may not be successful with its tender it will serve as an indicative outline of how the Contractor would carry out the service should it be successful.

The TSC provides, as an optional statement, for the contractor to submit a first plan for acceptance within a specified number of weeks after the Contract Date.

(b) Pre-contract planning

This covers the planning considerations once the contract has been awarded. It is essential that this planning stage takes place prior to commencing services on the Affected Property.

The process enables the contractor to:

- establish a clear strategy for carrying out the services.
- comply with contract conditions and requirements.
- establish a sequence on which the plan may be based.
- identify key dates by which parts of the services are required to be completed or the client is to provide information or resources to enable the contractor to complete services to the programme requirements.
- enable the assessment of contract budgets and cash flow forecasts
- schedule key dates for procurement of key materials and subcontractor tendering, placing orders and commencement.
- schedule pre-contract meetings.

(c) Service planning

The planning which is required to be implemented in order to maintain control and ensure that the services are completed on time and within the cost limits established at the tender stage.

The process enables the contractor to:

- monitor and update the master plan on a regular basis.
- optimise and continually review resources.
- keep progress under review and report on variances.

(iii) What is programming?

Programming is drawing all the planning considerations together into a presentable graphical format such as a bar chart and supporting tables which show the various operations, their durations, relationship between the operations, critical path and float, and the required resources so that it can be used as a reporting and control mechanism, and form the basis of progress reporting to the client.

The programme is an essential element of any contract, but while other contracts often pay only a passing interest to the programme, the NEC4 family has always seen it as a vital tool from which to:

(i) at a tender stage, assess the contractor's ability to carry out the service.
(ii) measure and report the contractor's actual progress against that planned.

(iii) assess the effect of compensation events upon completion and price.
(iv) ensure the Service Manager takes appropriate action when required.

Preparation of a plan

The process of writing a programme normally consists of three stages:

(i) Identifying the various operations to be carried out:

The operations may initially be written as a list, identifying each operation in chronological order. That order may depend on natural construction sequence, constraints to access and resource availability

(ii) Considering the duration of each operation

The duration of each operation can only be a forecast rather than an exact science as various issues have to be considered such as length of working days, differing productivity outputs, changes, resource availability, working conditions, changing weather and equipment breakdowns which can affect forecast durations. The contractor will then make appropriate adjustments as the work progresses.

(iii) The relationship between each operation will involve inserting links between operations which have a dependency on each other e.g. "start to start", "finish to start", "finish to finish". From this can be established what the logical path is in order to execute the various operations, operations need to start at the same time, what operations have to be completed before another operation can start, and what operations must finish at the same time.

The TSC does not prescribe the specific form of the plan or programme to be submitted for acceptance must take, it would normally be a bar chart programme. But the requirements of Clause 31.2 require much more information than just a Gantt chart.

Before considering in detail the requirements of the contract, it is worth reviewing other forms of programme in common use.

Bar charts

Any programme, at any stage of a tender or project, is normally written as a bar chart or Gantt Chart (named after Henry Laurence Gantt, 1861–1919, a US engineer) showing operations which may occur either on or off site including their sequence and duration, the start and finish of the bar depicting the start and finish date if the operation. Linked bar charts are the usual form of programme in common use on construction contracts. They are often supplemented by other information.

ADVANTAGES OF BAR CHARTS

*   The simple format allows them to be readily constructed, they are user friendly and understood by most practitioners.
*   They illustrate the specific activities, their durations and during the progress of the service, actual versus planned progress can be clearly shown.

- They can be used at all stages of the planning process through pre-tender, pre-contract and contract planning. They assist in showing the relationship between the pre-tender programme, the master programme and short term programmes.
- They clearly relate to the service delivery sequence, the use of linking on bar charts aids the overlapping of related activities and can enable the critical path to be identified.
- They are easily updated, at weekly and monthly time intervals for review purposes and progress reports.
- Key milestone symbols may be introduced to highlight critical dates with regard to key contract stages and information requirements.
- Resources may be readily related to the rate of working: labour histograms, value forecasts (value/time), and cumulative labour and plant forecasts in the form of service budgets.
- The programme forms the basis of financial forecasting for both the client and the contractor as a cash flow forecast can be prepared by pricing the programme.

DISADVANTAGES OF BAR CHARTS

- Without the use of links, they cannot properly depict the relationship and dependency between activities; for example where an activity must be wholly or partially complete before the next can start.
- Complex interrelationships cannot always be clearly shown.
- They may give too simple a picture, which may be misleading.

Elemental trend analysis (line of balance) charts

The elemental trend analysis (line of balance), as with many programming methods, originated from the manufacturing industry and is used where there is repetition of service either in number of units such as housing, or in repetitive floors in constructing or refurbishing multi-storey buildings. The unit or floor numbers are shown on the vertical y-axis, and the timeline of the project on the horizontal x-axis.

ADVANTAGES OF ELEMENTAL TREND ANALYSIS CHARTS

- They are very suitable for repetitive work and strict trade sequencing as the flow of work can be clearly demonstrated.
- They are more sophisticated than bar charts and simpler than network diagrams.
- They clearly show the rate of progress of each trade.

DISADVANTAGES OF ELEMENTAL TREND ANALYSIS CHARTS

They are totally unsuitable for non-repetitive projects and largely unsuitable for limited repetition.

- They are not easily understood, user friendly or readily understood by all practitioners.

Project network techniques (precedence diagrams, network analysis, critical path analysis)

Project network techniques comprise a number of methods but all are related to the dependency and precedence of one activity over another.

Networks may be represented as arrow diagrams where activities are represented by arrows and events (start and finishes) by "nodes". Activities take time, and are represented as months, weeks, days or even hours, but events do not.

As stated earlier, the bar chart is a clear and simple method of plotting activities, but it has a clear disadvantage in that the relationship between activities cannot be properly represented. With networks, the relationship is immediately apparent and when a delay occurs specifically affected activities will stand out.

A project network illustrates the relationships between activities (or tasks) in the project, showing the activities either on arrows or nodes.

PREPARING NETWORK DIAGRAMS

The first operation as with any method is to compile a list of operations to be carried out.

One must then consider against each operations:

1.   Which operations must be completed before this operation can start?
2.   Which operations cannot start till this operation is completed?

NB: One should always consider when an operation can be done, rather than when one thinks it will be done.

ADVANTAGES OF NETWORK DIAGRAMS

1.   Extremely effective when one is considering highly critical operations with strict dependency of several operations upon each other.
2.   Float and critical path can be calculated fairly easily from such diagrams.

DISADVANTAGES OF NETWORK DIAGRAMS

1.   Require a great deal of time and effort to produce them.
2.   They are not practical for "normal" service operations, which do not require highly detailed planning and monitoring of progress.
3.   Do not provide a ready format for financial reporting.

Service contracts are typically managed using software which combines high quality presentation with ease of revision and updating. This software can also link the master programme to short term programmes and also resource schedules and cash flow forecasts. However, one must never forget the power of pen and paper.

3.2   The traditional approach of service contracts

The value of a fully comprehensive service delivery programme is often misunderstood and undervalued by practitioners, this is often exacerbated by the minimal express requirement for a programme in many standard forms of contract.

Various commonly used forms merely state that the contractor "submits the master programme for the execution of the services", "states the order in which it proposes to carry out the services" or "shall submit a detailed time programme. . . . .and shall submit a revised programme whenever the previous programme is inconsistent with actual progress or with the contractor's obligations" often without going into any detail as to the required content and structure of the programme, the timing of the submission, the timing of the submission of any revised programme or any formal acceptance of the contractor's programme.

The danger is that by understating the requirement for a programme within the contract, the true value of a programme is then misunderstood by the parties to that contract, and the programme submission and any acceptance can then descend into a "box ticking" exercise.

Clearly, although the programme requirement within these contracts is not spelt out in detail, it is advisable in any contract for the client to require the contractor to submit a programme, even though there may be no contractual requirement to do so, and it is also advisable for the contractor to submit a detailed programme even if there is no contractual requirement to do so. The submitted programme should also identify the requirements to be met by the client.

The programme must be a dynamic tool which addresses all aspects of the service including information and other requirements, and the procurement process and lead in times for sub-contractors and suppliers in addition to the service operations. It must be flexible, and reviewed and if necessary revised on a regular basis as the work progresses and various events and changes come to light including the adjustments to float and time risk allowances.

3.3   What does the TSC require?

Programme vs Plan

Most contracts, for example the NEC4 Engineering and Construction Contract (ECC) use the term "programme" in terms of what the contractor has to submit to show its intentions, in terms of timing, resources and methodology.

However, if we compare the ECC with the TSC:

(i)   With the ECC there is a logical sequence of operations that the Contractor must carry out in order to complete the project, while sequences under the TSC can vary, as there is often a combination of planned and reactive operations, and in addition Task Orders may be introduced.

(ii) The ECC is used for constructing a project within a specified time, which may be extended, but the TSC is for providing a service for a period of time which is fixed and cannot be changed in the traditional sense, owing to delays on the critical path, though there is provision (Secondary Option X23) for the Client, with the agreement of the Contractor, to extend the Service Period.

The TSC has therefore always used the term "plan", as a plan is a more strategic or high level document than a programme which is at a more detailed level.

With the TSC, which could have a Service Period of say ten years, the Contractor has a multitude of different operations to carry out over that Service Period, some specifically on a regular basis, some on a performance related basis, some on a reactive (as required) basis, so a detailed programme as if the Contractor was building a defined project would not be practical.

The TSC requires the Contractor to state in detail what it is planning to do, and to submit that plan to the Service Manager for acceptance. The Service Manager then either accepts the plan, or gives reasons for not accepting it.

The clear plan requirement enables it to be used to:

(i) assess the Contractor's ability to carry out the service in terms of the intended resources and within the contractual time frame.
(ii) demonstrate that all of the contract needs and requirements have been considered.
(iii) show the Contractor's actual progress against that planned.
(iv) show the effect of Task Orders and compensation events upon completion of the service and the price.

Dates in the contract

There are a number of dates in the TSC which require some definition:

(i) Contract Date

The Contract Date is the date when the contract came into existence, however this is achieved, for example by the Client issuing a letter of acceptance to the Contractor. Note that there is no pro forma contract agreement within the contract, the principle being that parties will enter into contracts in various ways, using their own pro formas, standard letters or exchanges of correspondence.

(ii) Starting date

The starting date is decided and defined by the Client in Contract Data Part 1, together with the Service Period.

The Contractor starts the service from the starting date, but does not have access to the Affected Property, and cannot physically start work on the Affected Property until the Client gives access, which is required to be given as stated in the Scope, or as required for a Task. It also carries risks from the starting date and throughout the Service Period, so has to provide the insurance certificates as stated in the contract, prior to the starting date.

(iii) Service Period

The Service Period is stated in Contract Data Part 1, and is the period that the Contractor Provides the Service, as stated within the Scope.

Key Dates

Unlike the ECC, the TSC does not provide for Key Dates as pre-defined dates when a stated Condition must be met. These Key Dates and Conditions could then be included within Contract Data Part 1.

The addition of Key Dates, for instance via a Z clause, should be considered by TSC users as they are often appropriate to term service contracts as there may be certain constraints on when certain services have to be provided, e.g. during holiday periods and plant or facility shutdowns.

If the Service Manager then decides that the Contractor has not met the Condition stated by the Key Date, and the Client incurs additional cost on the same property as a result of that failure, then the Contractor will be liable to pay that amount.

There could also be a compensation event if the Service Manager gives an instruction to change a Key Date.

Alternatively, the conditions and timings required could be provided in the service level table as set out in secondary option X17. Failure to provide the service to the level stated, which can include time, leads to pre-agreed damages being due.

3.4 The Contractor's plan

The TSC has quite comprehensive and specific requirements for a plan to be submitted by the Contractor, these requirements being stated in the seven bullet points within Clause 31.2.

A plan submitted for acceptance must be submitted in the form stated in the Scope.

Some practitioners say that the requirement within Clause 31.2 is very onerous and in some cases excessive. However, it is critical that the plan shows in a clear and transparent fashion what the Contractor is planning to do in Providing the Service over the Service Period, when and how long each operation will last, and what it needs from the Client and Others to be able to comply with it.

There is nothing within the requirements of Clause 31.2 that an experienced Contractor would not ordinarily include within a professionally constructed plan submitted for the benefit of itself and the receiving party. Whether and how it chooses to show the information could be another matter.

There are two alternatives for the Contractor's submission of its first plan for acceptance:

(i) It may include, or be required to include, a plan with its tender in which case it is referenced by the Contractor in Contract Data Part 2. Note that under Clause 11.1(1) *"the Accepted Plan is the plan identified in the Contract Data or is the latest plan accepted by the Service Manager"*. Therefore, any plan referred to in Contract Data Part 2 automatically becomes the first Accepted Plan even though it may not necessarily comply fully with the requirements of Clause 31.2.

(ii) If a plan is not identified in Contract Data Part 2, the Contractor submits a first plan to the Service Manager for acceptance within the period of time after the Contract Date, this period of time being stated in Contract Data Part 1.

Such is the importance of a plan within the TSC that if no plan is identified in Contract Data Part 2, one quarter of the Price for Service Provided to Date is retained (Clause 50.5) until the Contractor has submitted the first plan showing the information the contract requires.

Note that a Contractor who submits a first plan which shows the required information, but the Service Manager does not accept it, would not be liable for withholding of payment under this clause. The clause relates to the Contractor's failure to submit a plan showing the information the contract requires, not the acceptance of it.

Note that when using Option A (Priced Contract with Price List), the Contractor must provide information showing how each item in the Price List related to the operations on the Plan and the Task Order programme it submits for acceptance.

Clause 31.2

The Contractor shows on each plan which is submitted for acceptance:

• *the starting date and the end of the Service Period*

These dates are stated by the Client in Contract Data Part 1.

- *for each Task*
  - *the Task starting date*
  - *planned Task completion and*
  - *the Task Completion Date*

- *the order and timing of the work of the Client or Others as last agreed with them by the Contractor or, if not so agreed, as stated in the Scope.*

The Contractor will state what it requires the Client or Others to do in order that the plan can be met. These dates may already have been agreed with the Contractor, but may also have been inserted into the Scope by the Client. "Others" are parties outside the contract, i.e. not the Client, Service Manager, Adjudicator, the Contractor or a Subcontractor or Supplier to the Contractor. Examples of "Others" would be regulatory authorities, other contractors working for the Client, utilities companies, etc.

This requirement is often provided in the form of an "Information Required" spreadsheet identifying what information is required and the latest date for providing it, failing which a compensation event can occur (Clause 60.1(6)).

It is important to recognise that the plan is not just a bar chart but may include spreadsheets, schedules and graphs.

- *provisions for:*

  - *time risk allowances*

Time risk allowances are essentially a form of float and are included within the duration of a specific operation. Essentially, a time risk allowance is the difference between what is the most "optimistic" (shortest) duration to complete an operation, and what is a "realistic" (expected) duration, bearing in mind the risks that the Contractor may face in completing that operation. It is important that the Contractor's staff are fully informed as to how the time risk allowances are being presented, so there is no misunderstanding of how the works have been planned and priced.

The Contractor's time risk allowances are to be shown on the plan but there are no requirements in the contract as to how these allowances should be shown, three possible methods being:

(i) Show a single bar for an operation, within which the time risk allowance is shown. This is difficult to do with some planning software.
(ii) Show a separate bar which represents the time risk allowance for a specific operation. However, this adds to the lines on the plan.
(iii) Show a separate column identifying the time risk allowance included within the total duration of the operation.

Clause 63.9 states that "*if the assessment of the effect of a compensation event is made using Defined Cost, it includes risk allowances for cost and time for matters which have a significant chance of occurring and are not compensation events*". It follows that they should be retained in the assessment of any delay, owing to the effect of a compensation event. They should be clearly identified as such in the plan or included in the time periods allocated to specific activities.

Example of a time risk allowance

The Contractor has an operation in its plan for "carrying out external surveys". It believes that if the resources are available, the weather is reasonable, and there are no problems with the surveying equipment, then the external surveys will take three weeks to complete.

However, knowledge and experience tells the Contractor that, with this type of service, rarely is everything as it plans and almost inevitably it will encounter some delay, so a realistic time scale would be four weeks. It should then show a four-week duration for the external surveys, within which one week is the time risk allowance.

It is important to recognise that time risk allowances belong to the Contractor and are retained in the assessment of any delay owing to a compensation event.

– health and safety requirements and

This will include time taken in complying with contractual and statutory requirements, but also in inducting new staff and operatives.

– the procedures set out in the contract

This can include time scales for acceptance of design, etc.

- *the dates when, in order to Provide the Service in accordance with the plan, the Contractor will need*
  - *access to the Affected Property as stated in the Scope or required for a Task;*
  - *Acceptances;*
  - *Plant and Materials, equipment and other things to be provided by the Client; and*
  - *information from Others*

Generally, the Contractor should be given access to the Affected Property, acceptances, anything provided by the Client, or information from Others to allow it to Provide the Service efficiently and without delays.

However, these are all matters which are outside the Contractor's control, so it is important that the Contractor clearly states in advance within the plan when it requires these items.

This information may be highlighted as a milestone on the bar chart itself, or attached to it in the form of information required schedules, resource schedules, etc.

Anything submitted by the Contractor as part of the plan must clearly and simply show the relevant operations for which access, Plant and Materials or information are required, and the implications if they are not provided.

It is important that the Contractor and the Service Manager then manage the process by raising any concerns in respect of their provision as early warnings.

- *for each operation, a statement of how the Contractor plans to do the work identifying the principal Equipment and other resources which will be used*

The Contractor is required to provide a statement of how it intends to carry out each and every item on each plan submitted for acceptance.

While accepting that the contract requires this information, this can be a somewhat idealistic clause which requires the Contractor to comply with the daunting and in many cases somewhat unreasonable requirement to produce all the information in its first plan submitted for acceptance. Particularly when providing a large or complex service with a long duration the Contractor may not yet know how it will carry out each operation and what resources it will use. It also may not have appointed a subcontractor for a part of the works so will not have method statements from the company who is actually carrying out the works.

It is also a significant task for the Service Manager to accept all the information the Contractor has submitted, so it may be appropriate in this respect, for the Contractor and the Service Manager to agree a rolling plan of submission of method statements as the work progresses and the methods and resources may become clearer, perhaps each submission covering the next three or four months operations.

Method statements have become increasingly important in industry particularly as health and safety legislation makes the parties more responsible for stating the methodology they are going to use in carrying out the works, and how they have provided for the various risks involved. For a plan to be meaningful it should always be linked to method statements so that the methodology if carrying out the works and also the resources the Contractor intends to use are clearly stated.

The contract does not specify the form these statements should take and in many cases a Contractor will spend a great deal of time producing detailed statements, which can be almost meaningless. The quality requirements in Section 4 of the contract offer some more description of what is required from the Contractor in this regard. They may be presented as a written statement or in tabular form but they should clearly show how the Contractor intends to do the work, main headings being:

  (i)   Title of each operation
 (ii)   Description of method
(iii)   Quantities (where relevant)
 (iv)   Principal equipment to be used with outputs and durations
  (v)   People type and team sizes with outputs and durations
 (vi)   Plant to be installed
(vii)   Materials to be used

NB: Cost is not normally included in method statements.

When preparing method statements, the Contractor should always consider and show:

  (i)   What is to be done
 (ii)   How it is going to be done
(iii)   Who is going to do it
 (iv)   Where it is to be done
  (v)   When it is to be done

This information will then initially allow the Service Manager to consider whether the Contractor is providing sufficient resources to Provide the Service within the time available and also in a safe manner, but also as the plan is revised and methods of delivery and resources used may change the Service Manager is then kept up to date as to the Contractor's intentions.

This is particularly important in potentially hazardous environments such as airports, factories, railways, etc.

This information will also prove invaluable when assessing compensation events.

* *other information which the Scope requires the Contractor to show on a plan submitted for acceptance.*

Again, this shows that the Scope requires the Contractor to show on a plan submitted for acceptance is not just the bar chart, but information on all that the Contractor is required to do in Providing the Service.

From these requirements once can see that the plan under an TSC contract is not just a bar chart, but a collection of documents including method statements, risk assessments, resource analyses, etc. All of these are submitted by the Contractor as its plan, and all of these must be considered by the Service Manager in determining whether to accept the plan. Extreme care should be taken to only require documentation which has a purpose and use in the administration.

Changed method of working (Option A only)

Clause 55.2 of Option A requires that if the Contractor changes its method of working so that the item descriptions on the Price List do not relate to the operations on the Accepted Plan, it submits a revision of the Price List to the Service Manager for acceptance.

Under Clause 55.3, a reason for not accepting a revision to the Price List is that:

* it does not relate to the operations on the Accepted Plan or a Task Order programme;
* any changed Prices are not reasonably distributed between the items in the Price List; or
* the total of the Prices is changed

The Service Manager's response

Clause 31.3

Within two weeks of the Contractor submitting a plan for acceptance, the Service Manager either accepts the plan or notifies the Contractor of the reasons for not accepting it.

A reason for not accepting a plan is that:

* *the Contractor's plans which it shows are not practicable*

The Contractor could be planning to use a particular working method or piece of Equipment that the Service Manager believes would not be appropriate? Or the Contractor has allowed a particular quantity of people or a planned output, which the Service Manager believes is not practicable.

* *it does not show the information which the contract requires*

It may not show a Task starting date or time risk allowances

* *it does not represent the Contractor's plans realistically*

The Service Manager may not believe the Contractor can complete an operation within the time it has allowed, or it could be showing planned completion of a Task, which the Service Manager believes is too early and therefore unrealistic?

- *it does not comply with the Scope.*

The Contractor may not have allowed for a constraint within the Scope. Or the plan may simply be in the wrong format. For example the Scope may prescribe that the plan has to be compiled and submitted using a particular brand of software, but the Contractor has used a different brand?

It has to be said that in many cases the Service Manager does not accept because the Contractor has misunderstood the contract requirements, it has not included all the stated constraints, or has not taken proper account of all the factors included in the Scope.

It is important that the Service Manager either accepts or does not accept the Contractor's plan and if it does not accept it, it should be clearly stated why not.

It has been known for Service Managers to respond to the submission of the plan either by making no comment at all within the two weeks they have to respond, or by stating that the plan is accepted on condition that the Contractor makes certain amendments to it.

Note that under Clause 31.3, if the Service Manager does not notify acceptance or non-acceptance of the plan within the time allowed, the Contractor may notify of that failure. If the failure continues for a further one week after the Contractor's notification, then it is treated as acceptance of the plan.

This provision within the NEC4 contracts brings some clarity to the process as previously Contractors stated that they were uncertain of the status of their plan or programme when they received no response.

Revised plans

The timing for submission of revised plans for acceptance is set out in Clause 32.2:

- *within the period for reply after the Service Manager has instructed the Contractor to*

For example, the Service Manager could instruct the Contractor to submit a plan to enable an issue to be discussed in an early warning meeting?

- *within the period for reply after the Service Manager has instructed a change to the Affected Property*
- *when the Contractor chooses to, and in any case. . .*

Again the Contractor may feel it beneficial to submit a plan to be discussed in an early warning meeting.

- *. . . and in any case at no longer interval than the interval stated in the Contract Data throughout the Service Period*

This is the longest period for submission of a revised plan
The revised plan shows:

- *the actual progress achieved and its effect upon the timing of the remaining work and services*

The progress achieved may be shown on the plan itself, or appended to it in the form of a progress report. Operations which are not yet completed will normally be shown as percentage complete, being the progress achieved to date against the total work within the operation on the date the assessment is made. It is useful if the Contractor and the Service Manager agree the progress statement before the revised plan comes into being. The submitted revised plan is then an undisputed statement of fact.

The revised plan should reflect planned v actual progress including early or delayed start, early or delayed completion of each operation.

It is important that, if the Contractor states a percentage completed to date, it should show actual progress achieved rather than time expired to date. Also, the Contractor should not just make statements such as "operation on plan" as this may be misleading. An operation on plan does not necessarily mean that the operation is not delayed, for example the Contractor is expecting to provide an operation late, then stating the words "operation on plan" can be interpreted as either in accordance with the contract, or progressing late in accordance with its expectations.

It is also important to recognise that many operations may not progress on a "straight line" basis, for example, undertaking electrical tests of 10,000 items may involve some work at heights, or during closures, overnight possessions, etc..

- *how the Contractor plans to deal with any delays and to correct notified Defects*

The Contractor should show in its revised plan any delays which it may or may not have caused, and also time needed to correct notified Defects.

- *any other changes which the Contractor proposes to make to the Accepted Plan*

An example of this could be that the Contractor re-sequences part of the work or proposes to use a different method or different Equipment to that specified in a previous method statement.

It is therefore very much a "living" plan. If the Contractor does not submit a plan which the contract requires, then the Service Manager must assess compensation events, without receiving a quotation from the Contractor (Clauses 64.1 and 64.2).

The Accepted Plan:

- effectively provides an agreed record of the progress of the service and where the delays have come from;
- provides a realistic base for future planning by both the Contractor and Service Manager; and
- is the base from which additional costs are derived.

Because each operation has a method statement and resources attached to it, the change in resources can be calculated and hence the change in costs. Because information is so much more transparent, there is more scope for working together.

3.5   The Task Order programme

Clause 33.1

The Contractor is required to submit a Task Order programme to the Service Manager for acceptance within the period stated in the Contract Data.

The Contractor shows on each Task Order programme is submitted for acceptance (see above for similar contents within the submission of the Contractor's plan):

- *the Task starting date and the Task Completion Date;*
- *planned Task completion;*
- *the order and timing of the work of the operations which the Contractor plans to do in order to complete the Task; and*
- *provisions for*

  – *float*

Float is any spare time within the Contractor's programme, after time risk allowances have been included, and represents the amount of time that operations may be delayed without delaying following operations (free float) and/or planned Completion (total float). It can also represent the time between when the Contractor plans to complete and when the contract requires it to complete (terminal float).

Float absorbs to a certain extent the Contractor's own delays or the delays caused by a compensation event, thereby lessening or avoiding any delay to planned Completion. In effect, no delay arises unless float on the relevant and critical operations reduces to below zero.

Programming is never an exact science, so float gives some flexibility to the Contractor in respect of incorrect forecasts or its own inefficiencies. As the work progresses the float will change as output rates change, Contractor's risk events take place and also compensation events arise.

The general belief, certainly among many Contractors, is that float belongs to them, as they wrote the programme, and therefore have the right to work to that programme and use any float that it contains. Conversely, Clients believe that if it is free time then they have the right to use it, but it actually depends where the float is and what it is for.

In principle float other than terminal float or time risk allowance is a shared resource, it is spare time, it belongs to the project, and therefore may be used by whichever party needs it first.

There are three primary types of float:

 (i) the amount of time that an operation can be delayed before it delays the earliest start of following operation (free float);
 (ii) the amount of time that an operation can be delayed before it delays the earliest completion of the works (total float);
(iii) the amount of time between planned Completion and the Completion Date (terminal float).

Generally, with other contracts, if the Contractor shows that it plans to complete a project early, then it is prevented from completing as early as planned, but it still completes before the contract completion date, then although there may be an entitlement to an extension of time under the contract, for example, adverse weather, none will be awarded as there is no delay to contract completion, but it may have a right to financial recovery subject to it proving loss and/or expense.

The TSC deals with this in a different way with Task Orders. If the Contractor shows on its programme planned Task Completion earlier than the task Completion Date and is prevented from completing by the planned Task Completion Date by a compensation event, then, when assessing the compensation event, Clause 63.6 states "*a delay to a Task Completion Date is assessed as the length of time that, owing to the compensation event, planned Task Completion is later than planned Task Completion as shown on the Task Order Programme current at the dividing date*", therefore any terminal float is retained by the Contractor, the period of delay being added to the Task Completion Date to determine the change to the Task Completion Date.

Any delay to planned Task Completion owing to a compensation event therefore results in the same delay to the Task Completion Date. Therefore, the extension is granted on the basis of time the Contractor is delayed, i.e. entitlement, not on how long is needed to achieve the current Task Completion Date.

- *time risk allowances;*
- *health and safety requirements;*
- *the procedures set out in the contract;*

- *the dates when, in order to Provide the Service in accordance with the Task Order programme, the Contractor will need:*
  - *access to the Affected Property;*
  - *Acceptances;*
  - *Plant and Materials, equipment and other things to be provided by the Client; and*
  - *information from Others.*
  - *for each operation, a statement of how the Contractor plans to do the work identifying the principal Equipment and other resources which will be used; and*
  - *other information which the Scope requires the Contractor to show on a Task Order programme submitted for acceptance.*

The Service Manager's response

Clause 33.3

Within one week of the Contractor submitting a Task Order programme for acceptance, the Service Manager notifies the Contractor of the acceptance of the Task Order programme, or reasons for not accepting it.

A reason for not accepting a Task Order programme is that:

- *the Contractor's plans which it shows are not practicable;*
- *it does not show the information which the contract requires;*
- *it does not represent the Contractor's plans realistically; and*
- *it does not comply with the Scope.*

Again, as stated above with the Contractor submission of the plan, it is important that the Service Manager either accepts or does not accept the Task Order programme, and if it does not accept it gives reasons why not.

Under Clause 33.3, if the Service Manager does not notify acceptance or non-acceptance of the Task Order programme within the time allowed, the Contractor may notify of that failure. If the failure continues for a further week after the Contractor's notification, then it is treated as acceptance of the Task Order programme.

Revised Task Order programme

The revised Task Order programme shows:

- *the actual progress achieved on each operation and its effect upon the timing of the remaining work*
- *how the Contractor plans to deal with any delays and to correct notified Defects*

The Contractor should show in the revised plan any delays which may or may not be caused by them, and also time needed to correct the Defects.

* *any other changes which the Contractor proposes to make to the Task Order programme*

The Accepted Task Order programme:

* effectively provides an agreed record of the progress of the services and where the delays have come from;
* provides a realistic base for future planning by both the Contractor and Service Manager;
* is the base from which changes to the Task Completion Date are calculated; and
* is the base from which additional costs are derived.

Because each operation has a method statement and resources attached to it, the change in resources can be calculated and hence the change in costs. Because information is so much more transparent, there is more scope for working together.

The timing for submission of revised Task Order programmes for acceptance is defined by Clause 34.2:

* *within the period for reply after the Service Manager has instructed the Contractor to*
* *when the Contractor chooses to*

3.6   Acceleration

Note that there is no provision within the TSC for instructing acceleration to bring forward completion of the service within a plan or a Task Order programme, the reason being that the Contractor Provides the Service over the duration of the Service Period and completes Tasks in accordance with the Task Order programme.

If a TSC user wishes to include acceleration provisions, then they must do so by drafting appropriate Z clauses.

3.7   Record-keeping

One of the most important, but often understated elements of managing a contract and its progress and any difficulties, whether one is a Contractor or a Service Manager, is the keeping of concise and timely records. It is beneficial if the records are agreed by the Contractor and the Service Manager. These are then undisputed records of fact. It is then only the interpretation of these facts which might subsequently be at issue. The use of asset management systems on service contracts tends to support good record keeping. The addition of mobile devices such as tablets has further increased the convenience of such systems in recent years.

A good record keeping system will enable the Contractor or the Service Manager to notify early warnings promptly, to assist the preparation of plans and Task Order programmes, to enable payments to be made under Options C and E, and also to prove invaluable in pricing Task Orders and in assessing compensation events, so readily available records are a must.

Records can include:

* Diaries
* Meeting Minutes

- Photographs and videos, including dates
- People returns
- Equipment returns
- Weather records
- Correspondence, electronic and physical
- Financial accounts and records

Also, on the rare occasions when a dispute arises and is referred to adjudication or the tribunal, records are invaluable.

4 Quality Management

4.1 Introduction

Quality Management is covered within the TSC by Clauses 40.1 to 44.2 and additional clauses in Main Options C and E.

Quality management is a major management function within the built environment industry. Unless a contractor can guarantee its client a quality service, it cannot compete effectively with others in the modern market.

Quality often stands alongside cost as a major factor in contractor selection by clients. To be competitive and to sustain good business prospects, quality systems need to be designed and maintained which are efficient and evidential.

The role of quality management for a contractor is not an isolated activity, but intertwined with all the operational and managerial processes of that company.

The modern concept of quality is considered to have evolved through three major stages over many years.

These stages are as follows:

(i) Quality control

The earliest and most basic form of quality management is quality control, the term being defined by an interpretation of its two elements – "quality" and "control".

Quality

The term "quality" is often used to describe prestige products such as expensive jewellery and motor cars. However, although applicable to these items, the term "quality" does not necessarily refer to prestigious products but merely to the fitness of the product or service to the customer's requirements. Quality is therefore described as meeting the requirements of the customer. In TSC contracts, those requirements are stated in the Scope.

Control

The concept of being "in control", or having something "under control", is readily understood, we mean we know what we intend to happen, and we are confident that we can ensure that it does.

Quality control introduces inspection to stages in providing services, ensuring that they are undertaken to specified requirements. The quality is measured by comparing the work actually carried out against the standard specified.

DOI: 10.1201/9781003463771-5

Inspection is the process of checking that what is produced is what is required. It is about the identification and early correction of defects.

The major objectives of quality control can be defined as follows:

(i) to ensure the completed service meets the specification
(ii) to reduce clients' complaints
(iii) to improve the reliability of services delivered
(iv) to increase clients' confidence
(v) to reduce delivery costs

Quality control is primarily concerned with defect detection and correction. The main quality control technique is that of inspection and statistical quality control techniques (i.e. sampling) to ensure that the services delivered and the materials used are within the tolerances specified. Some of these limits are left to the inspector's judgement and this can be a source of difficulty.

Controlling quality

The central feature to all quality control systems is that of inspection. To be effective the service delivery process requires that services to be inspected must be catalogued into a quality schedule.

The outcome of an inspection can be viewed in both objective and subjective ways:

(i) Objective

That which is quantifiable and measurable – line, levels, verticality and dimensions.

There are some precise quantified inspections including the testing of plant and machinery, pressure tests in pipework and tests on electrical installations.

(ii) Subjective

That which is open to the inspector's interpretation, e.g. cleanliness, fit, tolerances and visual checks.

Quality control implemented in service delivery

Traditionally there is one document that is used to determine the required quality of a service, this is the specification, which is referred to as the Scope in the TSC. The Contractor and Service Manager use the Scope during the service period to facilitate "quality" service delivery.

Many quality checks are undertaken visually although, particularly with plant installations, apparatus and semi-automatic monitoring are also prevalent.

It would be much better if the Contractor's managers and operatives had a clear understanding of the quality required, were able to recognise it themselves, and able to achieve it first time or regulate it by self-inspection. This concept being the basis of quality assurance potentially reduces the risks of producing unsatisfactory work and being involved in expensive re-work. Increasingly, self-certification is becoming common in the industry backed up only by random sampling checks by clients.

Notwithstanding the existence of self-certification as part of quality assurance and the emergence of total quality management most clients still engage monitoring staff to reassure themselves. However, the impact and importance of the clients' monitoring staff is much reduced in a quality assured or total quality managed company.

(ii) Quality assurance

Whereas quality control focuses on defect detection and correction when the service is being delivered, quality assurance is based on defect prevention.

Quality assurance is the process of ensuring that standards are consistently met thereby preventing defects from occurring in the first place.

"Fit for purpose" and "right first time, every time" are the principles of quality assurance.

Quality assurance concentrates on the delivery methods and procedural approaches to ensure that quality is built into the system.

Quality assurance is described under the following headings:

* evolution of quality assurance from quality control;
* definition of quality terms;
* quality assurance standards;
* developing and implementing a QA system; and
* quality assurance in construction.

(iii) Total quality management

This is based on the philosophy of continuously improving goods or services.

The key factor is that everyone in the company should be involved and committed from the top to the bottom of the organisation.

The successful total quality managed company ensures that their goods and services can meet the following criteria:

 (i) Be fit for purpose on a consistently reliable basis
 (ii) Delight the customer with the service which accompanies the supply of goods
(iii) Supply a quality service that is so much better than the competition that customers want it regardless of price

It may be argued that successful service companies have to meet at least two of these criteria to stay successful. The pursuit of total quality is seen as a never-ending journey of continuous improvement.

4.2   The Scope

Let us now consider what the TSC requires in terms of Quality Management, first the Scope.

The Contractor is required to Provide the Service in accordance with the Scope.

Note that there is no hierarchy or precedence between documents within the Scope and in fact within the contract. If there is any ambiguity or inconsistency in or between the documents which form the contract, the Service Manager or Contractor notifies the other, and the Service Manager the states how the inconsistency or ambiguity should be resolved (Clause 17.1).

If the Scope, a Task or the Affected Property has to be changed then this is a compensation event (60.1(1)) which is assessed as if the Prices and the Task Completion Date were for the interpretation most favourable to the party which did not provide the Scope (see Clause 63.11).

The Scope will normally be comprised of the following:

(i) Description of the Service

The Scope must include a clear definition of what is expected of the Contractor in Providing the Service. It must cover all aspects of the service, either specifically or by implication, otherwise it may be deemed to be excluded from the contract.

A comprehensive description should therefore be considered, which should be supplemented by specific detailed requirements in the form of a specification. If reliance is placed solely on the description of the service, there is a danger that an item may be missed and could be the subject of later contention.

While the Scope must clearly describe the boundaries for the service to be undertaken by the Contractor, it may also describe the Client's objectives and explain why the work is being undertaken and how it is intended to be used. It says what is to be done and what is to be included in general terms, but not how to do it.

The Scope is the document that a tenderer can look to, to gain a broad understanding of the scale and complexity of the service and be able to judge its capacity as a Contractor to undertake it. It is written specifically for each contract and should be comprehensive, but it should be made clear that it is not intended to include all the detail, which will also be contained in the drawings, specifications and schedules.

(ii) Drawings

Where appropriate, drawings should detail the work to be carried out by the Contractor and be complementary to the other constituents of the Scope.

(iii) Specification

The Specification is a written technical description which describes the character and quality of services, materials and workmanship, for work to be executed. It may also lay down the sequence in which various portions of the services are to be executed. As far as possible, it should describe the outcomes required, rather than how to achieve them, criteria or standards which are required for the work, and complementary to the other constituents of the Scope and any notes or other text on the drawings, and the provisions of the contract.

(iv) Responsibilities of the parties

The responsibilities of the contracting parties and their representatives should be clearly defined in a manner which will leave no doubt as to the obligations that each is accepting, while also considering the written terms of the conditions of contract. This will also allow the procedures necessary to enable the contract to progress satisfactorily from inception to completion to be established at the outset of the service.

The contract as a whole will also define the responsibilities for the planning of the services, timings, resources, the supply of any free issue materials, the issuing of instructions and the form which these instructions are to take, the programming of the services, the method of measuring and evaluating the services, the circumstances which will constitute a change to the scope of works and the duties of the parties during service delivery.

4.3   Defining testing within the Scope

The main documents within the Scope in respect of testing and inspection will refer to what needs testing and inspecting, by whom and by when.

Testing can be related to individual services within the Affected Property and factory testing outside the Affected Property.

While the Scope will set out the procedures and parameters for testing and commissioning, the contract will set out the Parties' contractual obligations, and the consequences of failure to meet them. Many activities, for example electrical testing, will have to comply with legislation independent of the contract.

The documentation required in the form of test certificates, warranties and guarantees, to enable proper records to be maintained should be stipulated, as will the effect in relation to the guarantee or warranty period and the extent to which it will apply. Many such records are only maintained electronically so access to the relevant asset management systems should be specified for the client together with the requirements for data transfer at expiry of the service period.

Liability for defects should be defined in the contract together with the period of time for which such liability is to apply and the length of time within which defects are to be remedied.

It is imperative that the Scope, produced at tender stage and identified in Contract Data Part 1, defines the full extent and timing of tests and inspections required and the subsequent correction of any defects.

This is very often not covered in sufficient detail within the Scope, so it is important for the compiler of the Scope to consider a number of key questions:

(i)   What is the purpose of the test?

What is it intended to prove or disprove? Is the test required before a following service activity can be carried out by the Contractor or another party, or is it required before the Client can use a building or part of the building?

(ii)   What is to be tested?

What elements or components are required to be tested? Is it to be tested in an assembled or unassembled state?

(iii)   How is the test to be carried out, and who is to provide the materials facilities and equipment to do the test

Whether the test is a well-known "industry standard" test, or not the exact nature of the test must be clearly defined. Who supplies the equipment, who pays for the power or fuel for the equipment, etc? It is also important to detail who is to provide the materials and facilities to do the test and who meets the costs.

(iv)   Who should do the test? Who should be present and who should be invited should they wish to attend when the test is taking place?

There may be a requirement for an independent party to carry out the test and provide results.

(v) When and where will the test take place? On site or off site?

Clearly state at what stage or on what date the test must take place. Sometimes, tests may have to be carried out on a periodic basis, for example on concrete samples on delivery and as the concrete cures.

(vi) What is the expected result or outcome from the test?

What are we expecting the test to prove? What results do we need to gather once the test has been carried out?

(vii) Who should be advised of the outcome of the test?

The Service Manager will normally be advised of the outcome of a test, but sometimes a sub-contractor, or a third party outside the contract may need to be informed.

(viii) What should be done in the event of a successful test e.g. certification?

Does a certificate have to be issued following successful completion of a test? Does that successful outcome mean that the Contractor can proceed to the next stage of the service? Is payment to the Contractor dependent on successful completion of the test?

(ix) What should be done in the event of an unsuccessful test e.g. rectification or replacement, and further testing?

Does the Contractor have to repeat the test? Is the retest done immediately after the first test, or should there be a period before the next test? Is it possible that even though something has failed a test, it may still be considered acceptable, possibly by the Contractor offering a cost saving?

(x) What about additional testing i.e. to confirm a suspected defect or the extent of a suspected defect?

This may be referred to as a "search" where the Contractor may be instructed to "open up" work which is believed to be defective, possibly including the requirement for some form of additional testing.

There may be a requirement for Plant and Materials, provided under a Task Order, to be tested off site before delivery.

4.3 Quality management system

Clause 40.1 requires the Contractor to operate a quality management system as stated in the Scope, submitted to the Service Manager for acceptance, and complying with the requirements stated in the Scope.

A quality management system is a set of interrelated or interacting elements that companies and other organizations use to direct, manage and control how quality policies are implemented and quality objectives are achieved.

A process-based quality management system uses a process approach to manage and control how its quality policy is implemented and quality objectives are achieved.

The quality management system will normally be to ISO9000 standards, ISO9000 defining a Quality Management System as "co-ordinated activities to direct and control an organization with regard to the degree to which a set of inherent characteristics fulfills the requirements"

ISO 9001 sets out the basic requirements for a quality management system, generally that it enables a company to consistently provide products or services that enhance customer satisfaction while meeting applicable statutory and regulatory requirements. It is essentially about providing quality assurance to customers.

Quality management principles

The standard is based on a number of quality management principles including a strong customer focus, the motivation and implication of top management, the process approach and continual improvement. Using ISO 9001:2015 helps ensure that customers get consistent, good quality products and services, which in turn brings many business benefits.

A quality management system is intended to help a company to:

* identify, develop and implement efficient management systems;
* reduce waste to acceptable levels;
* ensure effective team working; and
* regularly audit and review the Quality Management system to identify excellence, problems and areas for improvement.

Promoting customer loyalty by:

* ensuring customers' needs are fully identifies and understood;
* ensuring project requirements are identified, understood, documented and agreed by all;
* providing a management system that ensures on time delivery of the agreed product or service;
* providing brief, user friendly and easily accessible methods to express satisfaction or dissatisfaction and ensure that any dissatisfaction is resolved; and
* meeting the national statutory and regulatory requirements.

To provide a good working environment and culture for staff by:

* promoting a culture of honesty, timely communication and assistance to each other;
* providing brief, user friendly and easily accessible procedures and processes that reflect the users' preferred method of working wherever practicable; and
* ensuring that recognition is given to deserving staff.

Audits

Checking that the system works is a vital part of ISO 9001:2015. An organisation must perform internal audits to check how its quality management system is working. An organisation may decide to invite an independent certification body to verify that it is in conformity to the standard, but there is no requirement for this. Alternatively, it might invite its Clients to audit the quality system for themselves.

4.4 Quality policy statement

A Contractor's quality policy statement is required by Clause 40.2 to be submitted by the Contractor to the Service Manager for acceptance. The quality policy statement defines its management's commitment to achieving and maintaining quality. It is often just a single paragraph, but it should clearly state the Contractor's general quality orientation and clarify its basic intentions.

Quality policy statements should be used to generate quality objectives and should serve as a general framework for action. Quality policies can be based on the ISO 9000 Quality Management Principles and should be consistent with the organization's other policies.

A typical quality policy statement could say:

"ABC Contractors is committed to achieving and consistently exceeding its Client's expectations, getting it right first time every time, by providing services to the highest quality and within a safe and environmentally friendly manner. Our success in achieving these objectives is continuously measured through our quality management system accredited under ISO 9001:2015".

4.5 Quality plan

The Contractor is also required by Clause 40.2 to submit a quality plan to the Service Manager for acceptance. The quality plan is a document that is used to specify what procedures and processes will be followed, and who is responsible for managing them, to ensure that the quality expectations of the customer will be met.

This may consist of the following sections:

- The Client's quality expectations – What the contract requires
- Acceptance criteria – How quality will be measured
- Quality responsibilities – Who is responsible
- References to any quality standards – What other codes of practice, legislative and/or other standards apply
- Quality controls and audit processes – How the procedures and processes will be checked for compliance
- Change management and procedures – How changes will be incorporated into the plan

The quality plan should be then signed by all relevant parties to confirm that they have read and fully understand their responsibilities.

The degree to which the Service Manager checks conformity with the statement and plan is entirely a matter for the Client. However, under Clause 40.3, if it discovers a non-compliance the Contractor is required to correct such failure on the instruction of the Service Manager. Such instruction would not be a compensation event. The Service Manager assesses the compliance of a quality plan submitted for acceptance.

4.6 Testing and inspections

The Contractor and the Service Manager are required to inform each other of tests and inspections before they start, and also the results of those tests. If a test shows that any work has a Defect, the Defect is corrected and the test repeated. The Service Manager assesses the cost incurred by the Client in repeating a test or inspection after a Defect is found and the Contractor pays the amount assessed.

The Service Manager is also obliged to do the tests and inspections without causing unnecessary delay to the work, or to a payment which is conditional upon a successful test. The term "unnecessary delay" is subjective, clearly it must be expected that there may be some delay, whilst the Service Manager does tests and inspections. If the test or inspection causes unnecessary delay, then it is a compensation event (Clause 60.1(10)).

4.7   Defining a Defect

A Defect is defined under Clause 11.2(5) as:

* *a part of the service which is not in accordance with the Scope; or*
* *a part of the service which is not in accordance with the applicable law or the Accepted Plan.*

Note that it is considerably more difficult to identify a Defect in a Contractor's <u>services</u> than it is within a Contractor's <u>works,</u> as it is less visible, also the Defect may not become apparent for some time!

The Scope provides the reference point for what the Contractor has to do to Provide the Service. If a part of the services provided by the Contractor does not comply with the Scope, then the Contractor is obliged to correct the Defect so that it does comply with the Scope.

In most cases of Defects the work done will fall short of the Scope, but as the words "not in accordance with" are used, in theory, work which exceeds the Scope would also be a Defect. Clearly, the Client is likely to accept such a Defect under Clause 44!

A Defect may not necessarily mean that the services provided are not fit for their intended purpose, just that they are not in accordance with the Scope.

It is also possible that the services provided are "defective" but as they comply with the Scope there is no Defect!

4.8   Notifying and correcting Defects

It is important to stress that the Contractor is responsible for meeting the quality standards required by the contract and has an obligation to correct Defects whether or not the Service Manager has notified it of them. In that respect it should adopt a quality assurance/quality control system which prevents Defects occurring and corrects them promptly if they do occur.

Under Clause 43.1, until the end of the Service Period, the Service Manager and the Contractor are required to notify each other as soon as they become aware of a Defect (See example template Figure 4.1)

It is perhaps curious that the Contractor has to notify the Service Manager of each Defect when it becomes aware of it, when the Service Manager probably does not need to know about each Defect and if it did, it surely wants to know that the Contractor has corrected the Defect, rather than merely became aware of it!

Note that the clause is referring to Defects in the services, not in the works (for example, under a Task Order) that have been carried out by a Contractor. If a Contractor has carried out defective work then it should be fairly apparent what it has to do to correct the Defect, i.e. replace something, rebuild it, or carry out some form of remedial work to it to make good.

However, a Contractor having to correct a Defect raises a number of complex issues, for example, if the Contractor has been carrying out a survey and has missed an item from the survey, it is a Defect, but how can the Client calculate the costs of having the Defect corrected by other people? One assumes that the Client calculates the amount in a fairly arbitrary manner and the Contractor pays this as a form of damages, without anyone actually correcting it?

The Client's rights in respect of a Defect which the Client has not found or notified by the end of the Service Period are not affected. This refers to "latent defects" which become apparent after the Service Period has ended and should apply to any latent defect, not just one that the Client has not found or notified.

Under Clause 43.2, the Contractor corrects a Defect whether or not the Service Manager notifies them of it.

Under Clause 43.3, the Contractor has to correct Defects within a time which minimises the adverse effect on the Client or Others. It is difficult to ascertain a time that minimises an adverse effect, but the Clause goes on to say that If the Contractor does not correct a Defect within the time required by this contract, the Service Manager assesses the cost to the Client of having the Defect corrected by other people and the Contractor pays this amount, so the reference to "the time required by this contract" is presumably that time which minimise the effect of the Defect.

Problems that are not a Defect (i.e. under this contract) may occur for which the Contractor is not liable (see earlier note under definition of Defect). Nevertheless, the Contractor must correct the Defect, but it is a compensation event under Clause 60.1(12).

4.9   Liability for the cost of correcting Defects

The Contractor's liability for the cost of correcting Defects varies according to the Main Option used. Note that under Options C and E where the Contractor is reimbursed its Defined Cost, Disallowed Cost is only for the cost of *"correcting Defects caused by the Contractor not complying with a constraint on how it is to Provide the Service stated in the Scope"*.

However, that provision may not apply to all Defects in Providing the Service, for example something could be "defective" but it was not a Defect *"caused by the Contractor not complying with a constraint on how it is to Provide the Service stated in the Scope"*, in which case the Contractor may be paid for correcting some Defects.

While some may see this as the Contractor's right to repeatedly attempt to get the work right at the Client's expense it must be remembered that particularly in the case of Option C this is a target contract, therefore if the Contractor is paid for correcting a Defect, not only is this bad for its reputation, but as it is not a compensation event it is not being given additional time to correct the Defect and the additional cost paid will reduce the entitlement to Contractor's share at Completion.

It may be appropriate for clarity to define and include further Disallowed Cost provisions as Z clauses (see Chapter 5).

4.10   Accepting Defects

In the event that a Defect becomes apparent, there are two choices:

(i) The Contractor corrects the Defect so that the part of the service is in accordance with the Scope, the applicable law or the Accepted Plan where applicable.

(ii) The Contractor and the Service Manager may each propose to the other that the Scope should be changed so that the Defect does not have to be corrected (Clause 44). Note that "each proposes to the other" requires the acceptance of the Contractor and the Service Manager, the latter probably discussing the matter with the Client.

Example: the Contractor has carried out a large area of replacement wall tiling in a maintenance contract for a large hotel. While visually the quality of the work appears to be very good, the tiles have not been laid to the required tolerance within the Scope and therefore the work is defective. The hotel rooms are due to be completed and re-opened in two weeks, so if the Contractor has to take down the wall tiles, re-order another batch of tiles and then lay the new tiles to the required tolerance that could take a considerable time and possibly prevent completion of the service or that specific part of the service.

In this case, provided both the Contractor and the Service Manager are prepared to consider changing the Scope, then the Contractor provides a quotation to the Service Manager stating a financial saving (reduced Prices), based on it not having to correct the Defect and/or an earlier Task Completion Date, and if the Service Manager accepts the quotation then the Scope is changed and the Prices and/or Task Completion Date are changed in accordance with the quotation.

This is obviously not an option where the service is not in accordance with the applicable law.

If the Service Manager accepts the quotation it gives an instruction to change the Scope, the Task, the Prices and the Task Completion Date as applicable.

4.11   Defects arising after the Service Period

Unlike the Engineering and Construction Contract (ECC), there is no Defects Certificate under the TSC.

This is somewhat curious as the Contractor is carrying out a service over the Service Period, which may include building or engineering work, provision of materials, etc. for which defects in workmanship and/or materials may arise. In addition the Contractor may be liable for providing a faulty service.

For that reason, many Clients include a Defects Certificate within the TSC so that any outstanding Defects are brought to the Contractor's attention and are signed off as they are completed.

Note that if there is a Defects Certificate it is not conclusive in that if a Defects Certificate states that there are no Defects, it does not prevent the Client from its rights should a Defect arise later, or if the Service Manager did not find or notify it.

The Client will normally only list what are usually defined as "Patent Defects", i.e. those which are observable from reasonable inspection at the time, and may not include what are usually defined as "Latent Defects", which may be hidden from reasonable inspection and may come to light at a later date, examples being some structural Defects.

The Contractor's and Others' liability for correction of latent defects and other costs associated with them will be dependent on the applicable law, and liability may remain despite the issue of a Defects Certificate.

Most countries have a statutory limitation period for latent defects which come to light at a later date, beyond which a claim cannot be made for defective work.

In England and Wales, for example, the time limit for actions founded on simple contract is six years and for a specialty contract or deed, the time limit is twelve years, in both cases this is measured from the date on which the cause of action accrued, normally with the TSC, this is taken as the end of the Service Period.

5 Payment

5.1 Introduction

Payment, including assessment, certification and payment of amounts due are covered within the TSC by Clauses 50.1 to 53.4, with further references within the Main Option Clauses and, where appropriate, Secondary Option Y(UK)2.

5.2 The Main Option Payment Clauses

There are separate payment provisions within the TSC, dependent on the Main Option chosen, each Option referring to two main terms:

(i) "The Prices"

These are the various elements that make up the total price for carrying out the service and will be in the form of a Price List and/or target (Option C), dependent on the Main Option selected.

(ii) "The Price for Service Provided to Date"

This term is used in making the assessment of amounts due to the Contractor, and is again dependent on the Main Option chosen:

Definition of the Price for Service Provided to Date within Options A, C and E

Option A	• The Price for each lump sum item in the Price List which the Contractor has completed; and • Where a quantity is stated for an item in the Price List, an amount calculated by multiplying the quantity which the Contractor has completed by the rate
Option C	• The total Defined Cost which the Service Manager forecasts will have been paid by the Contractor before the next assessment date plus the Fee
Option E	• The total Defined Cost which the Service Manager forecasts will have been paid by the Contractor before the next assessment date plus the Fee

Option A: Priced contract with Price List

Let us first consider in more detail the terms "Prices" and "The Price for Service Provided to Date".

*   *The Prices are the amounts stated in the Price column of the Price List. Where a quantity is stated for an item in the Price List, the Price is calculated by multiplying the quantity by the rate.*

DOI: 10.1201/9781003463771-6

- *The Price for Service Provided to Date is:*
  - *the Price for each lump sum item in the Price List which the Contractor has completed; and*
  - *where a quantity is stated for an item in the Price List, an amount calculated by multiplying the quantity which the Contractor has completed by the rate*

It is important to note that the Price List is the basis for the Contractor to be paid, it is not Scope or a schedule of what the Contractor has to do, or is intending to do, in Providing the Service.

Example

A Contractor has been appointed to carry out a property survey of 50 houses for a local authority Client, the work being programmed to take four months. The survey will be carried out in Months 1 and 2, and the compilation and submission of results being carried out in Months 3 and 4.

That part of its Price List which covers the property survey could then look like this:

Complete property surveys

Survey Houses 1–10	£8,500
Survey Houses 11–20	£8,850
Survey Houses 21–30	£8,500
Survey Houses 31–40	£7,950
Survey Houses 41–50	£8,750
Compilation of results – Houses 1–25	£5,000
Compilation of results – Houses 26–50	£5,000

In this case, it is vitally important that the Contractor and the Service Manager fully understand what constitutes completion of a survey, so that there are no debates between them about when an item in the Price List is completed.

Completion of an item in the Price List means that the work has no notified Defects which will delay the work of the Contractor, the Client or Others.

So, in the example above, if the Contractor has completed the field survey to Houses 11–20, but there is a Defect within the survey that needs to be corrected, then as long as the Defect would not delay the work of the Contractor, the Client or Others, then the item in the Price List is complete.

Note that if several operations are grouped under one item on the Price List, the Contractor is not entitled to be paid for that item until the whole item is completed, it is "all or nothing".

Assessing the Contractor's right to payment under Option A is therefore different to the traditional approach of valuing the works carried out to date by measuring it.

Note that, if the effect of a compensation event is to delay the completion of an item in the Price List, the Contractor is not entitled to be paid a proportion of the price for that item, the delayed payment having to be priced within the quotation for the compensation event that caused the delay.

The Contractor must show how each item in the Price List relates to operations on the Contractor's plan and each Task Order programme, and if, possibly owing to a change of method of working it does not, the Contractor must submit a revision to the Price List to the Service Manager for acceptance (Clause 55.2).

If a Contractor has establishment costs such as mobilising, providing accommodation, time-based preliminaries and similar activities as well as physical operations, it is important that the items in the Price List are clearly defined not only in terms of what the work is, but where it is, and when it will be completed, so that their completion for payment purposes is not in doubt.

It is important where the service provided by the Contractor includes preliminaries items, to correctly allocate these items within the Price List, as they could contribute as much as 10 per cent to 15 per cent of the total of the Prices.

Preliminaries could normally comprise the following:

(i) Fixed Preliminaries

These would comprise preliminaries which may be single items, not directly influenced by progress within the Service Areas.

Typical examples are – establishing a depot and offices, delivery of temporary accommodation and other equipment to Site, connection of telephones and other temporary services.

These should be identified within the Price List as single activities e.g. "installation of site accommodation".

(ii) Time-Related Preliminaries

These comprise preliminaries which are based on time working within the Service Areas, rather than having any direct relationship to the quantity of work carried out.

Typical examples are –management salaries, site accommodation hire charges, etc

These should be identified within the Price List as time-based items e.g. "one month hire of temporary accommodation". Again, when the items are completed, they are included within the next assessment.

(iii) Value-Related Preliminaries

These would consist of preliminaries which are directly related to how much work is being carried out at any particular time.

Typical examples are mechanical equipment, e.g. power tools, etc.

These, which are not always referred to as preliminaries should be included within the operations on which they are being used rather than as stand-alone items in the Price List.

Option C: Target contract with Price List

The definition of the Prices is the same as Option A:

* *The Prices are the amounts stated in the Price column of the Price List. Where a quantity is stated for an item in the Price List, the Price is calculated by multiplying the quantity by the rate.*

However, the definition of the Price for Service Provided to Date is different:

* *The Price for Service Provided to Date is the total Defined Cost which the Service Manager forecasts will have been paid by the Contractor before the next assessment date plus the Fee.*

The payment is then based on a combination of Defined Cost paid, and also yet to be paid by the Contractor. For example, if the current assessment date is 1 October, then Defined Cost, which is forecast to be paid by the next assessment date on 1 November, is included.

While the Service Manager's forecast may appear to be no more than a subjective guess, the Contractor is required to have an up-to-date plan and Task Order programme, and also it has probably received many of the invoices which it will be paying before the next assessment date, but they are yet to be paid. These will feature in the Defined Cost forecast that the Contractor must provide to the Service Manager on a regular basis (Main Options C and E Clause 20.4)

Many practitioners question the role of the Price List under Option C, when unlike Option A, it is not used for assessing payments. The answer is that the Price List under Option C assists the Service Manager and Client in assessing tenders as they can see how the Contractor has priced its tender. The Price List is then updated throughout the period of the contract.

Note that the Contractor under Option C is also required to advise the Service Manager on the practical implications of the Accepted Plan and on subcontracting arrangements.

The Contractor is required to prepare forecasts of the total Defined Cost for the whole of the service at intervals as stated in the Contract Data, from the starting date until the end of the Service Period, and submit them to the Service Manager.

Option E: Cost-reimbursable contract

The definition of the Prices is the same as Option A and C:

*   *The Prices are the amounts stated in the Price column of the Price List. Where a quantity is stated for an item in the Price List, the Price is calculated by multiplying the quantity by the rate.*

The definition of the Price for Service Provided to Date is the same as Option C:

*   *The Price for Service Provided to Date is the total Defined Cost which the Service Manager forecasts will have been paid by the Contractor before the next assessment date plus the Fee.*

The Contractor is again required to advise the Service Manager on the practical implications of the Accepted Plan and on subcontracting arrangements.

In addition, the Contractor is again required to prepare forecasts of the total Defined Cost for the whole of the service at intervals as stated in the Contract Data, from the starting date until the end of the Service Period, and submit them to the Service Manager.

The difference with Option E, is that there is no comparison and reconciliation against a target at the end of the Service Period.

5.3   Unfixed materials within or outside Service Areas

A common area of confusion if the Contractor is providing materials within a TSC is the payment to the Contractor for unfixed materials within or outside the Service Areas i.e. materials on or off site. Unlike other forms of contract, the TSC has no express provisions for payment for such unfixed materials.

In order to be paid for these materials the Contractor should consider the payment rules for each of the Main Options:

Option A

As the Price for Work Done to Date is based on lump sums or quantities within the Price List there may be provision already made within the Price List for such unfixed materials.

Option C and E

As the Price for Work Done to Date is the Defined Cost which the Service Manager forecasts will have been paid by the Contractor before the next assessment date plus the Fee, the Contractor must provide accounts and records to show that it has paid for the unfixed materials, or it will have paid for them by the next assessment date.

Note that, in respect of unfixed materials within the Service Areas, whatever title the Contractor has to Plant and Materials passes to the Client if it has been brought within the Service Areas and passes back to the Contractor if it is removed from the Service Areas with the Service Manager's permission (see Clause 70.1)

5.4 Accounts and records

Option C and E require the Contractor (Clause 52.2) to keep the following records:

- accounts of payments of Defined Cost;
- proof that the payments have been made;
- communications about and assessments of compensation events for Subcontractors; and
- other records as stated in the Scope.

The Contractor must also allow the Service Manager to inspect these records at any time within working hours, not just in assessing and certifying payments (Clause 52.3).

The level of checking of the Contractor's accounts and records is at the Service Manager's (and the Client's) discretion, some wish to examine each and every cost and its relevant back up document, some wish only to select certain costs at random, others wishing to only carry out only a cursory check of costs.

Some Service Managers say that if the Parties truly are acting in a spirit of mutual trust and co-operation as the contract requires Clause 10.2), then there should be no need for a detailed examination of the Contractor's accounts, though this probably needs a quantum leap of faith for many Service Managers and their Clients!

It is important to establish the format on which accounts and records are to be presented by the Contractor to the Client, particularly in respect of assessing payments, and this should be detailed in the Scope, the following alternative methods normally being employed:

(i) The Contractor submits a copy of all of its accounts and records to the Service Manager on a regular basis. In reality this is usually provided electronically, and the Service Manager will have 24 hours access.

or

(ii) The Contractor gives the Service access to its accounts and records and the Service Manager takes copies of all the records it requires. While not expressly stated within the contract, the Service Manager should not have to travel too far to inspect them. It has been known for accounts and records to be "available for inspection" in a different country to the Service Areas, the Contractor stating that they were available for inspection at any time during normal working hours! Again the location and availability of accounts and records could be detailed within the Scope.

or

(iii) The Contractor gives the Service Manager direct access to its online accounts and records through secure access, the Service Manager can then take copies of the records it requires. This is now the most common method.

5.5   The Contractor's share (Option C)

The Contractor's share percentage for each share range is entered into Contract Data Part 1, the principle being that the Contractor <u>receives</u> a share of any saving and <u>pays</u> a share of any excess when the Price for Service Provided to Date is compared with the total of the Prices (the target) at the Completion of the whole of the service and in the final payment.

At the dates stated in the Contract Data and when the final amount due is assessed, the Service Manager makes an assessment of the Contractor's share by applying the Contractor's share percentage to the difference between the Price for Service Provided to Date and the total of the Prices. A final assessment is made using the final Price for Service Provided to Date and the final total of the Prices.

While NEC practitioners often refer to the terms "pain" and "gain" in Option C, many practitioners are surprised that these terms do not actually exist within the contract! The Contractor pays or is paid the Contractor's share.

Example

The Contractor's share percentages and the share ranges have been entered into Contract Data Part 1 as follows:

Share range	Contractor's share percentage
less than 80%	50%
from 80% to 90%	40%
from 90% to 100%	30%
greater than 100%	50%

At Completion of the service:

The total of the Prices (the target) = £7,200,000

The Price for Service Provided to Date = £6,100,000

Difference = £1,100,000

£7,200,000

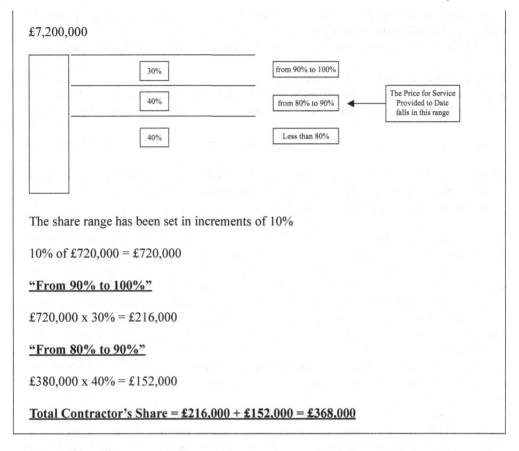

The share range has been set in increments of 10%

10% of £720,000 = £720,000

"From 90% to 100%"

£720,000 x 30% = £216,000

"From 80% to 90%"

£380,000 x 40% = £152,000

Total Contractor's Share = £216,000 + £152,000 = £368,000

5.6   Assessing the amount due

The first assessment date is decided by the Service Manager to suit the Parties, and will normally be based on the time that the Contractor has been Providing the Service, the Client's procedures and timing for processing and issuing payments, and the Contractor's payment requirements and internal accounting system.

Clearly, in this respect, some discussion needs to take place between the Service Manager, the Client and the Contractor in order that an assessment date convenient for all can be set.

The first assessment must be made within the "assessment interval" after the starting date, this is normally inserted in the Contract Data as "four weeks" or "one calendar month".

Later assessment dates occur at the end of each assessment interval until:

* four weeks after the end of the Service Period; or
* the Service Manager issues a termination certificate.

Note that there is no provision within the contract for a minimum certificate amount.

The Service Manager assesses the amount due at each assessment date, calculating the Price for Service Provided to Date using the rules of the specific Main Option.

The Contractor submits an application for payment to the Service Manager before each assessment date, and must include details of the application, the format to be in accordance with the Scope.

In making its assessment, the Service Manager considers "an application for payment the Contractor has submitted" before the assessment date.

If the Contractor submits an application for payment, the amount due to the Contractor is:

- the Price for Service Provided to Date,
- plus other amounts to be paid to the Contractor (e.g. Contractor's share, value added tax, etc.)
- less amounts to be paid by or retained from the Contractor (e.g. retention, delay damages)

If the Contractor does not submit an application for payment, the amount due is the lesser of:

- the amount the Service Manager assesses as due at the assessment date, as though the Contractor had submitted an application for payment before the assessment date; and
- the amount due at the previous assessment date.

Therefore, under the NEC4 Term Service Contract, if the Contractor does not submit an application for payment, the amount due would be ZERO, or could even be a minus figure.

Note that NEC3 and previous editions of NEC, used the phrase "any application", so there was no express requirement for the Contractor to submit an application.

It is essential that the Contractor either submits a plan with its tender or within the time scale specified within Contract Data Part 1. Failure to do so will entitle the Service Manager to retain one quarter of the Price for Service Provided to Date in its assessment of the amount due.

Note that the amount is only withheld if the Contractor has not submitted a plan which shows the information which the contract requires, e.g. method statement, time risk allowances, etc.

If the Contractor has submitted a plan which contains all the information that the contract requires, but the Service Manager disagrees with, for example, part of the method statement or the programme has not yet been accepted, then the provision, and the associated amount retained, does not apply.

Also, by using the words "if no plan is identified in the Contract Data" if the Contractor submitted a plan as part of its tender AND identified it in Contract Data Part 2, the provision would not apply, as the tender has been accepted and therefore the tender plan is part of that submission.

Note, also the additional Clauses C50.7 (Option C), E50.8 (Option E) and 50.9 Options C and E).

Under Clauses C50.7 (Option C) and E50.8 (Option E), payments of Defined Cost made by the Contractor in a currency other than the currency of the contract are included in the amount due as payments in the same currency, then such payments are converted to the currency of the contract in calculating Fee and any Contractor's share using the exchange rates.

Also, under Clause 50.9 (Options C and E), when a part of the service has been finalised, the Contractor notifies the Service Manager and makes the cost records available to demonstrate that it has bene correctly assessed.

The Service Manager then reviews the records and within 13 weeks either:

- accepts that part of Defined Cost as correct;
- notifies the Contractor that further records are needed; or
- notifies the Contractor of any errors.

If the Contractor provides further records, the Service Manager can within four weeks either accept that part of Defined Cost as correct, or notify the Contractor of the correct assessment.

If the Service Manager does not notify within four weeks, the Contractor's assessment is treated as correct.

This Clause means that the Contractor does not have to keep all the records of Defined Cost available for inspection throughout the Service Period, which would be quite a challenge if the Service Period was several years!

5.7 Payment

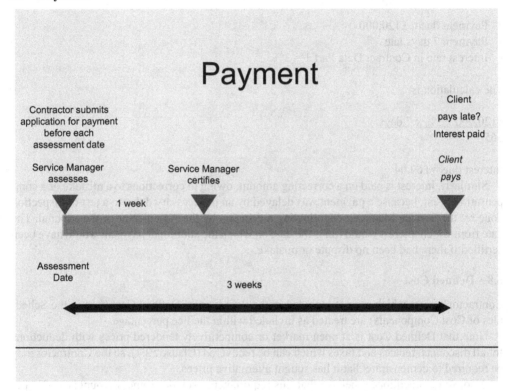

The Service Manager is required to certify the payment due to the Contractor within one week of each assessment date (note if the contract falls under the Housing Grants, Construction and Regeneration Act 1996, Secondary Option Y(UK) 2 applies).

The payment (by either Party) is made by the later of:

- one week after the paying Party receives an invoice from the other Party; and
- three weeks after the assessment date or, if a different period is stated in the Contract Data, within the period stated.

If the certificate is less than the previous one, then the Contractor pays the Client the amount due, in which case the Client sends an invoice to the Contractor.

Payments are in the currency of the contract unless otherwise stated.

Interest is paid if a payment is not made or the Service Manager does not issue a certificate which it should have issued. The interest rate is stated in Part 1 of the Contract Data and is assessed on a daily basis from the date the payment should have been made until the date when the late payment is made, calculated using the interest rate in Contract Data Part 1, and compounded annually.

The interest due can be calculated on the basis of the following formula:

Payment due x interest rate x the number of days late:
365
So if one assumes the following:

- Payment due = £120,000
- Payment 7 days late
- Interest rate in Contract Data Part 1 = 3%

The calculation is:

£120,000 x 3% x 7 days
365

Interest due = £69.04

Similarly, interest is paid on a correcting amount, owing to corrections to a mistake or a compensation event, because a payment was delayed by an unnecessary delay to a test or inspection done by the Service Manager, or following a decision by the Adjudicator or the tribunal. The date from which interest would run is the date on which the additional payment would have been certified if there had been no dispute or mistake.

5.8   Defined Cost

Contractor's costs which are not included in the definition of Defined Cost (within the Schedules of Cost Components) are treated as included within the Fee percentage.

Note that Defined Cost is at open market or competitively tendered prices with deductions for all discounts, rebates and taxes which can be recovered (Clause 52.1), so the Contractor may be required to demonstrate that it has sought alternative prices.

Also, some practitioners believe that this requirement excludes some discounts e.g. 2½ per cent 'cash' discount. However, the clause states "all discounts" so there are no exceptions.

5.9   Service Areas

The Service Areas are the Affected Property and those parts of the service areas which are:

- *necessary for Providing the Service; and*
- *used only to provide services in the contract.*

unless later changed in accordance with the contract   .

The Contractor may propose an addition to the Service areas within Contract Data Part 2. An example would be where the footprint of a building in which the service is to be provided fills the Service Areas and therefore the Contractor chooses to use an adjacent piece of land for its site offices and storage of materials.

In this case, it would identify the additional piece of land as an addition to the Service Areas within Contract Data Part 2.

In addition, under Clause 16.3, the Contractor may submit a proposal to the Service Manager for adding to the Service Areas during the carrying out of the service. It is essential that the additional area is necessary for Providing the Service and for no other reason.

The significance of the Service Areas is that Defined Cost is identified in the Schedule of Cost Components and Short Schedule of Cost Components as:

(i) The cost of employing people who are working within the Service Areas
(ii) The cost of Equipment used within the Service Areas
(iii) The cost of Plant and Materials including delivery to and removal from the Service Areas
(iv) The cost of Subcontractors
(v) The cost of charges paid by the Contractor for provision and use in the Service Areas
(vi) The cost of manufacture and fabrication of Plant and Materials which are designed specifically for the works and manufactured or fabricated outside the Service Areas
(vii) The cost of providing a shared service done outside the Service Areas

On Option A contracts, this is a consideration in pricing compensation events, on Options C and E it is a consideration in pricing compensation events, and also in assessing the Contractor's payments.

5.10   The Schedules of Cost Components

The TSC contains two Schedules of Cost Components:

* Schedule of Cost Components
* Short Schedule of Cost Components

and they have two uses:

* Under Option A, C and E they define the cost components, which are included in assessments of compensation events.
* Under Options C and E, they define the cost components for which the Contractor will be directly reimbursed.

Any other costs not included within the Schedule of Cost Components and Short Schedule of Cost Components including overheads and also profit, is deemed to be included in the Fee.

The "Schedule of Cost Components" or the "Short Schedule of Cost Components"?

As stated above, there are differing uses for Defined Cost, dependent on which main Option is chosen as follows:

Option A: Defined Cost is only used for assessing compensation events, therefore the Short Schedule of Cost Components is used

Option C and Option E: Defined Cost is used for assessing compensation events <u>and</u> for assessing Price for Service Provided to Date. Therefore the Schedule of Cost Components is used

(i) The Schedule of Cost Components

The Schedule of Cost Components only applies when Option C or E is used.

Clause 1: People

This relates to the cost of people who are directly employed by the Contractor and whose normal place of working is within the Service Areas, i.e. the Affected Property and any another area named in Contract Data Part 2, and also people whose normal place of working is not within the Service Areas but who are working in the Service Areas, i.e. people who are based outside the Service Areas, but who are working within the Service Areas temporarily.

The cost component covers the full cost of employing the people including wages and salaries paid while they are in the Service Areas, payments made to the people for bonuses, overtime, sickness and holiday pay, special allowances and also payments made in relation to people for travel, subsistence, relocation, medical costs, protective clothing, meeting the requirements of the law, a vehicle and safety training.

NEC4 contracts do not provide for rates for People to be priced as part of the Contractor's tender within the Schedule of Cost Components, the principle being that the People component is real cost rather than a rate that has been forecast at tender stage months or even years before it is required to be used. This is a fairer method of establishing cost in that the risk is not the Contractor's, but in practice requires the Contractor if necessary to prove the cost of each operative, tradesman and member of staff, which can at times be laborious.

For this reason, many Clients include a schedule of rates to be priced by the Contractor at tender stage, these rates the being used for payments and compensation events where appropriate. These rates are also referred to during the tender assessment process.

Note that within the January 2023 amendment, the third bullet point allows for People listed in the Contract Data who are employed by the Contractor, whose normal place of working is not within the Service Areas and who are working outside the Service Areas other than for manufacture and fabrication and providing a shared service.

This amendment could include for example a Contractor's commercial manager working from home, assuming they are working on, and costed to the specific contract.

Clause 2: Equipment

This relates to the cost of Equipment (see definition of this word in Clause 11.2(7)) used within the Service Areas. If the Equipment is used outside the Service Areas, then it is deemed to be included in the Fee percentage.

If the Equipment is hired externally by the Contractor, the hire rate is multiplied by the time the equipment is required. The cost of transport of Equipment to and from the Service Areas, and any erection and dismantling costs are also separately costed.

If the Equipment is owned by the Contractor, or hired by the Contractor from a company within the parent company such as an internal "plant hire" company, the cost is assessed at open market rates (not at the rate charged by the hirer) multiplied by the time for which the Equipment is required.

Equipment purchased for use in the contract is paid on the basis of its change in value (the difference between its purchase price and its sale price at the end of the period for which it is used) and the time-related on cost charge stated in Contract Data Part 2 for the period the Equipment is required.

During the course of the contract, the Contractor is paid the time-related charge per time period (per week/per month) and when the change in value is determined, a final payment is made in the next assessment.

Example

The Contractor has purchased 6 temporary office units @ £6,250 each for use on a project of four years duration. This cost includes supply and delivery of the units.

- Total purchase price $= 6 \, x \, £6,250 = £37,500$
- On completion, the units are sold for £2,750 each, therefore the total sale price is £16,500
- The change in value is therefore £37,500 $-$ £16,500 $=$ £21,000

Any special equipment is paid on the basis of its entry in Contract Data Part 2.

Consumables such as fuel are also separately costed including and materials used to construct or fabricate equipment.

The cost of transporting Equipment to and from the Service Areas and the erection, dismantling and any modifying of the Equipment is costed separately.

Any People cost such as equipment drivers and operatives involved with erection and dismantling of plant should be included in the cost of People, not the Equipment they work on.

Clause 3: Plant and Materials

This deals with purchasing Plant and Materials, including delivery, providing and removing packaging and any necessary samples and tests. The cost of disposal of Plant and Materials is credited.

Clause 4: Subcontractors

If work is sublet then the Cost Component includes the cost of Subcontractors excluding retention, or any other amounts retained from the Subcontractors, so that the Client does not pay or deduct the same amount twice.

Clause 5: Charges

This covers various miscellaneous costs incurred by the Contractor such as temporary water, gas and electricity supplies, payments to public authorities, and also payments for various other charges such as cancellation charges, buying or leasing of land, inspection certificates and facilities for visits to the Service Areas.

The cost of any consumables and equipment provided by the Contractor for the Service Manager's offices is also included as direct cost. (Note the cost of the Contractor's own equipment is covered in the Equipment Section.

Note that the Contractor's own consumables are classed as Equipment.

Clause 6: Manufacture and fabrication

This relates to the components of cost of manufacture or fabrication of Plant and Materials outside the Service Areas. Hourly rates are stated in Contract Data Part 2 for the categories of employees listed.

Clause 7: Shared services outside the Service Areas

If the Contractor provides people who are providing shared services outside the Service Areas, then these are paid by multiplying the applicable rates in Contract Data Part 2, by the time spent in providing the shared service.

This could be, for example, staffing of a call centre.

Clause 8: Insurance

The cost of events for which the Contractor is required to insure and other costs to be paid to the Contractor by insurers are <u>deducted</u> from cost.

(ii) The Short Schedule of Cost Components

The Short Schedule of Cost Components is restricted to the assessment of compensation events under Option A.

Clause 1: People

This relates to the cost of people who are directly employed by the Contractor and whose normal place of working is within the Service Areas, i.e. the site and any another area named in Contract Data Part 2, and also people whose normal place of working is not within the Service Areas but who are working in the Service Areas i.e. people who are based off site, but who are on site temporarily. It also includes people who are not directly employed by the Contractor but are paid according to the time worked while they are in the Service Areas e.g. security guards, cleaners, etc.

It also includes for the cost of people who are not directly employed by the Contractor but are paid according to the time worked while they are in the Service Areas. This may include for example security guards or cleaners who are paid by the hour.

Defined Cost is calculated by multiplying the applicable People Rates by the time spent on work within the contract.

Clause 2: Equipment

This relates to the cost of Equipment used within the Service Areas.

The cost of equipment is calculated by reference to a published list. In Contract Data Part 2, the Contractor names the published list to be used and also inserts a percentage for adjustment against items of equipment in the published list.

The Contractor also inserts rates into Contract Data Part 2 for Equipment not included within the published list. Any Equipment required which is not in the published list or priced within Contract Data Part 2 is then priced at competitively tendered market rates.

The time the Equipment is used is as referred to in the published list, which may have hourly, weekly or monthly rates.

By referring to the published list, whether the Equipment is then owned or hired by the Contractor is irrelevant.

The cost of transporting Equipment to and from the Service Areas and the erection and dismantling of the Equipment is costed separately. The cost of Equipment operators is included within the People costs.

Any Equipment not included within the published lists is priced at competitively tendered or open market rates.

Clause 3: Plant and Materials

This is exactly the same methodology as for the Schedule of Cost Components and also deals with purchasing, delivery, providing and removing packaging and any necessary samples and tests. The cost of disposal of Plant and Materials is credited.

Clause 4: Subcontractors

If work is sublet then the Cost Component includes the cost of Subcontractors excluding retention, or any other amounts retained from the Subcontractors, so that the Client does not pay or deduct the same amount twice.

Clause 5: Charges

This covers various miscellaneous costs incurred by the Contractor such as temporary water, gas and electricity, payments to public authorities, and also payments for various other charges such as cancellation charges, buying or leasing of land, inspection certificates and facilities for visits to the Service Areas.

The cost of any consumables and equipment provided by the Contractor for the Service Manager's offices is also included as direct cost. (Note the cost of the Contractor's own equipment is covered in the Equipment Section.)

Again, note that the Contractor's own consumables are classed as Equipment.

Clause 6: Manufacture and fabrication

This relates to the components of cost of manufacture or fabrication of Plant and Materials outside the Service Areas. The calculation is based on amounts paid by the Contractor.

Clause 7: Design

This is exactly the same as for the Schedule of Cost Components and deals with the cost of design outside the Service Areas. Again, hourly rates are stated in Contract Data Part 2 for the categories of employees listed.

Clause 8: Insurance

The cost of events for which the Contractor is required to insure and other costs to be paid to the Contractor by insurers are <u>deducted</u> from cost.

5.11 The Fee

While many practitioners believe that the Fee simply covers "overheads and profit", the definition is a little wider. All components of cost not listed in the Schedules of Cost Components are covered by the Fee percentage.

There is no schedule of items covered by the Fee percentage, but the following list, while not exhaustive, gives some examples of cost components not included within the Schedules of Cost Components:

   (i) Profit
  (ii) The cost of offices outside the Service Areas e.g. the Contractor's head office
 (iii) Insurance premiums
 (iv) Performance bond costs
  (v) Corporation tax
 (vi) Advertising and recruitment costs
(vii) Sureties and guarantees required for the contract
(viii) Some indirect payments to staff
 (ix) Specialist support staff at head office

From this one can see that there are some elements covered by the Fee percentage which could be mistakenly assumed as being covered by the Schedules of Cost Components.

5.12   Disallowed Cost

There are different clauses covering Disallowed Cost dependent on the selected Main Option:

Options C and E – Clauses C11.2.(24), E11.2.(24)

Disallowed Cost is cost which the Service Manager decides:

- *is not justified by the Contractor's accounts and records*

The Contractor is obliged to keep accounts, proof of payments, communications regarding payments, and any other records as stated in the Scope (Clauses 52.2 and 52.3). If it does not have the accounts and records to prove a cost then the cost must be disallowed. It is not sufficient to merely prove that the goods or materials are at the Affected Property for the Service Manager to see, or held at a designated place for inspection, as with many other contracts.

- *should not have been paid to a Subcontractor or supplier in accordance with its contract*

The Contractor is required to submit the name of each proposed Subcontractor to the Service Manager for acceptance. In addition the Contractor is obliged to submit the proposed subcontract documents, except any pricing information (see below), for each Subcontractor to the Service Manager for acceptance, unless an unamended NEC contract is proposed, or the Service Manager has agreed that no submission is required.

In addition under Options C and E, the Contractor is required to submit the proposed pricing information in the subcontract documents for each subcontract if the Service Manager instructs the Contractor to make the submission.

If the Contractor then pays a Subcontractor or supplier an amount that is not in accordance with its contract with them, then this is Disallowed Cost.

- *was incurred only because the Contractor did not*

(i) *follow an acceptance or procurement procedure stated in the Scope*

The Scope may require the Contractor to comply with a specific procedure in respect of, for example, submission of design proposals or procurement of Subcontractors. If the Contractor does not comply with the procedure then associated costs are disallowed.

(ii) *give an early warning which the contract required it to give*

If the Contractor incurs a cost that could have been avoided if the Contractor had given early warning, then it is disallowed.

(iii) *give notice to the Service Manager of the preparation for and conduct of an adjudication or proceedings of a tribunal between the Contractor and a Subcontractor or supplier*

and the cost of

• *Plant and Materials not used to Provide the Service (after allowing for reasonable wastage) unless resulting from a change to the Scope, a Task or the Affected Property*

Excess wastage of plant or materials beyond what is considered reasonable is a Disallowed Cost. The question of what is "reasonable" can often be debatable! As with all disallowed cost it is the Service Manager's responsibility to make the decision and to disallow the cost.

• *correcting Defects caused by the Contractor not complying with a constraint on how it is to Provide the Service stated in the Scope.*

The operative word within this bullet is "<u>how</u>". A constraint may be stated in the Scope, such as a prescribed method of working, timing or sequence to suit the Client's access or availability, or a particular equipment which has to be used.
   If the Contractor does not comply with a stated constraint and a Defect occurs, then the Contractor's cost of correcting the Defect is disallowed.

• *resources not used to Provide the Service (after allowing for reasonable availability and utilisation) or not taken away from the Service Area when the Service Manager requested.*

This provision will include People and Equipment. If the Contractor does not remove Equipment when it is no longer required, then this is disallowed cost.
   Note that if the Contractor is using more resources than it has planned and priced for, or its resources are inefficient, that is <u>not</u> a Disallowed Cost.

• *preparation for and conduct of an adjudication or proceedings of a tribunal between the Parties*

If an adjudication, arbitration or legal proceedings occur, then each party bears its own costs. This bullet was not introduced until NEC3 was published, so prior to NEC3 a Contractor could, in theory, refer a dispute to adjudication and whether or not it was successful, any costs arising could be included as cost and would not be disallowed. An agreement as to party costs made, such as here, prior to the service of a Notice of Adjudication would not be effective in British law.

5.13   Final assessment

The TSC does not have the equivalent of a Final Account or Final Certificate which is found in other contracts, certifying that the contract has fully and finally been complied with and that issues such as payments, compensation events and defects have all been dealt with.

However, NEC4 introduced a new provision where, in the case of the TSC, the Service Manager makes a final assessment of amounts due to the Contractor.

The Service Manager makes an assessment of the final amount due to the Contractor and certifies a payment, no later than:

- 13 weeks after the end of the Service Period or, if within a different period if stated in the Contract Data; or
- 13 weeks after the Service Manager issues a termination certificate.

Similar to a normal payment, the payment (by either Party) is made by the later of:

- one week after the paying Party receives an invoice from the other Party; and
- three weeks after the assessment date or, if a different period is stated in the Contract Data, within the period stated.

If the certificate is less than the previous one, then the Contractor pays the Client the amount due, in which case the Client sends an invoice to the Contractor.

If the Service Manager does not make the final assessment within the time allowed, the Contractor may issue to the <u>Client</u> an assessment of the final amount due. If the Client agrees with the assessment, the Party to which payment is due, submits an invoice to the other Party. Similar to a normal payment, the payment (by either Party) is made by the later of:

- one week after the paying Party receives an invoice from the other Party; and
- three weeks after the assessment date or, if a different period is stated in the Contract Data, within the period stated.

The assessment of the final amount due is conclusive, unless:
If Option W1 is selected, a Party:

- refers a dispute about the assessment of the final amount due to the Senior Representatives within four weeks of the assessment being issued;
- refers any issues not agreed by the Senior Representatives to the Adjudicator within three weeks of the list of issues not agreed being produced, or when it should have been issued; and
- refers to the tribunal its dissatisfaction with a decision of the Adjudicator regarding the final amount due within four weeks of the decision being made.

If Option W2 is selected, a Party:

- refers a dispute about the assessment of the final amount due to the Senior Representatives or to the Adjudicator within four weeks of the assessment being issued, but it may omit this stage by virtue of W2.2(1);
- refers any issues not agreed by the Senior Representatives to the Adjudicator within three weeks of the list of issues not agreed being produced, or when it should have been issued; and

- refers to the tribunal its dissatisfaction with a decision of the Adjudicator regarding the final amount due within four weeks of the decision being made.

If the above applies under Options W1 or W2, the assessment of the final amount due is changed to include:

- any agreement the Parties reach; and
- a decision of the Adjudicator which has not been referred to the tribunal within four weeks of the decision.

5.14   Project Bank Account (Option Y(UK)1)

As stated in an earlier Chapter, Option Y(UK)1 provides for a Project Bank Account to be set up which receives payments from the Client which is in turn used to make payments to the Contractor and Named Suppliers.

There is also a Trust Deed between the Client, the Contractor and Named Suppliers containing the necessary provisions for administering the Project Bank Account. This is executed before the first assessment date.

The Contractor should include in any subcontracts for Named Suppliers to become party to the Project Bank Account through a Trust Deed. The Contractor notifies the Named Suppliers of the details of the Project Bank Account and the arrangements for payment of amounts due under their contracts. The Named Suppliers will be named within the Contractor's tender, but also additional Named Suppliers may be included subject to the Client's acceptance by means of a Joining Deed, which is executed by the Client, the Contractor and the new Named Supplier. The new Named Supplier then becomes a party to the Trust Deed.

As the Project Bank Account is maintained by the Contractor, it pays any bank charges and also is entitled to any interest earned on the account. The Contractor is also required at tender stage to put forward its proposals for a suitable bank or other entity which can offer the arrangements required under the contract.

The process every month is that the Contractor submits an application for payment including details of amounts due to Named Suppliers in accordance with their contracts.

The Client makes payment to the Project Bank Account, the Contractor makes payment to the Project Bank Account of any amounts which the Client has notified the Contractor intends to withhold from the certified amount and which is required to make payment to Named Suppliers.

The Contractor then prepares the Authorisation, setting out the sums due to Named Suppliers. After signing the Payment Schedule, the Contractor submits it to the Client for signature and submission to the project bank.

The Contractor and Named Suppliers then receive payment from the Project Bank Account of the sums set out in the Payment Schedule after the Project Bank Account receives payment.

In the event of termination, no further payments are made into the Project Bank Account.

6 Compensation events

6.1 Introduction

Compensation events are covered within the TSC by Clauses 60.1 to 66.3.

Service contracts have to remain flexible to the needs of clients, end users and others and the fact that these needs may change as the contracts progress.

Service contracts also tend to last for several years and therefore they have to be able to accommodate change. Compensation events are much more than about change though, they also reflect the risk profile of the contract.

The Contractor may be compensated for a number of events, those events being for reasons which are at the Client's not the Contractor's risk.

When considering the principle of compensation events under the TSC, it is appropriate firstly to consider how change is managed with most other standard forms of contract.

The "traditional approach" to managing change

The "traditional approach" to managing change in service contracts is normally dealt with in the following way within the contract:

The changes in the form of additions, omissions or substitutions are priced by measuring the changed services, then pricing the change using rates consistent with the contractor's original tender, the pricing document for that tender normally being in the form of bills of quantities, a schedule of rates or a contract sum analysis.

Where the changed services are similar to those referred to in the pricing document, executed under similar conditions, and do not significantly change the quantity of work set out in the tender, the original rates and prices apply. Where the changed work is similar to that referred to in the tender, but not executed under similar conditions, and/or there is a significant change in the quantity of work, the original rates and prices form the basis of the pricing i.e. they are adjusted to allow for the differences in conditions and/or quantities. There are normally provisions for the parties to agree fair rates and prices and where the work cannot be clearly defined and/or measured then hourly rates or daywork provisions apply.

While this has always been seen as the "normal" way to price change in a service contract, it has its drawbacks.

The first problem with the traditional approach is that the rates and prices inserted by the contractor in its tender were calculated based on the original scope of work and whether they are correct or incorrect they are being applied to changes which are outside the Contractor's control. In that sense, either the Contractor or the Client may gain or lose by the changes.

DOI: 10.1201/9781003463771-7

The second problem is that there are rarely specific time scales within the contracts for submitting the proposed revised prices and agreeing them, so the agreement of the effect of a change can be prolonged, usually months, but sometimes years after the work is completed.

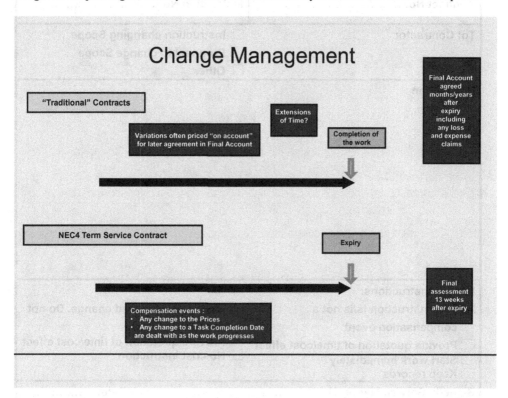

The NEC approach

The NEC contracts have always taken a very different approach to pricing and managing change as contained within the compensation event provisions, whereby the event is notified, priced, assessed and implemented within a defined, and relatively short time scale.

A very important difference is that what circumstances constitute the basis of a compensation event are defined within the contract. Importantly, this gives the contract drafter the facility to adjust the balance of risk between the client and the contractor.

The experience of the authors of this book are that often the requirements of the compensation events procedures are poorly understood and applied.

The onerous requirements on the Service Manager introduced within NEC to reply to notices and quotations were included to meet a particular failing as were the time bar requirements on both Parties.

6.2   What is a compensation event?

There is no definition of a compensation event stated within the TSC. However, it could be explained as "*an event which is a Client's risk, and if it occurs, and affects the Contractor,*

Contract: ………………………………..	**SERVICE MANAGER'S INSTRUCTION**
Contract No: …………………………….	Instruction No…………………………….

To: Contractor

☐ **Instruction changing Scope**
☐ **Proposal to change Scope**
☐ **Other**

Description

Further instructions:

☐ **This instruction is/is not a compensation event**
☐ **Provide quotation of time/cost effect**
☐ **Start work immediately**
☐ **Keep records**

☐ **Issued as proposed change. Do not implement.**
☐ **Provide quotation of time/cost effect**
☐ **No cost instruction**

Signed: (Service Manager) ……………………………………   **Date:** ……………

Contractor's Summary

☐ **Delay to a Task Completion Date is………..........days/weeks**
☐ **No delay to a Task Completion Date**
☐ **Change in Defined Cost plus Fee Plus/Minus £** …………………
☐ **No change in Defined Cost plus Fee**

Note: This is a summary only. Full supporting documentation or records must accompany this statement.

Signed: (Contractor) …………………………………….   **Date:**……………

Copied to:

Contractor ☐   Service Manager ☐   File ☐   Other ……………….. ☐

entitles the Contractor to be compensated for any effect the event has on the Prices and/or a Task Completion Date", note that unlike other NEC 4 contracts the Service Period does not get extended by a compensation event.

6.3 The compensation events

The compensation events may be found in a number of locations within the contract:

- 18 events are stated in sub-clauses 60.1(1) to (18); these are core clauses and therefore always apply, unless amended by Z clause provisions.

Note that sub-clause 60.1(18) is not a compensation event in itself, but provides for additional compensation events to be stated in Contract Data Part 1.

- Further events are included in Secondary Options X2 (Changes in the Law), X10 (Information Modelling), X12 (Multiparty Collaboration), X29 (Climate Change) and Y(UK)2 (The Housing Grants, Construction and Regeneration Act 1996)
- Compensation events can also be added, omitted or amended through Z clauses.

The principal compensation events are set out in Clause 60.1:

60.1(1) The Service Manager gives an instruction changing the Scope, a Task or the Affected Property except:

- *a change made in order to accept a Defect; or*
- *a change to the Scope provided by the Contractor for its plan which is made*

  - *at the Contractor's request; or*
  - *in order to comply with the Scope provided by the Client.*

Changes to the Scope are normally defined in other contracts as "Variations". The TSC does not use the term Variations, but the Service Manager can give an instruction changing the Scope, a Task or the Affected Property, which can be in the form of an addition, omission or alteration to the services.

An instruction can also be given by the Service Manager in order to resolve an ambiguity or an inconsistency within any of the documents that make up the contract (Clause 17.1), such instructions often being a change to the Scope. An ambiguity may be defined as "an uncertain meaning or intention" and therefore is something that is unclear; an inconsistency may be defined as a "conflict in meaning or intention" and therefore could be a mismatch or discrepancy between documents.

Example

An example of an inconsistency is found in a highways maintenance contract.

One part of the Scope, under the section "Safety Barriers", requires damage to central reservation barriers to be repaired within 4 hours of any road traffic damage being caused. Elsewhere, under the section "Safety Critical Structures", damage caused by road traffic has to be repaired in a time that minimises delay to road users.

> The Contractor notifies the Service Manager of the inconsistency under Clause 17.1.
> Having discussed the matter with the Client, the Service Manager instructs a change to the Scope. The second requirement is changed to a response time of four hours and the matter is resolved. On this occasion the Contractor did not notify a compensation event but in similar situations it might have done had the revised instruction created more onerous obligations.

A compensation event which arises from an instruction to change the Scope to resolve an ambiguity or inconsistency is assessed *"as if the Prices and the Task Completion Dates were for the interpretation most favourable to the Party which did not provide the Scope"* (Clause 63.11). In the example above it would be assessed in favour of the Contractor who is deemed to have priced the cheaper, easier or quicker alternative.

Note, from the first bullet point in Clause 60.1(1), if the change to the Scope is made in order to accept a Defect (Clause 44), then it is not a compensation event.

As detailed previously in Chapter 4, the Contractor and the Service Manager may each propose to the other that the Scope should be changed so that a Defect does not have to be corrected, in effect nullifying the Defect. This would not be a compensation event.

Also, from the second bullet point, a change to the Scope provided by the Contractor for its plan which is made either at its request or to comply with other Scope provided by the Client is not a compensation event.

So Scope provided by the Client in Part 1 of the Contract Data takes precedence over that provided by the Contractor in Part 2 of the Contract Data.

Thus it is critical that the Contractor ensures that any Scope it prepares and submits with its tender as Part 2 of the Contract Data, complies with the requirements of the Client's Scope in Part 1 of the Contract Data.

Example

Tendering processes often require tenderers (i.e. prospective Contractors) to state how they will perform services if they are successful in the competition. Typically these statements will form Scope submitted by the Contractor as part of its tender within Contract Data Part 2.

If the Scope in Contract Data Part 1 requires an electrical test of switchgear annually but the Contractor has stated that such tests will be done every two years then there is an inconsistency. This will be resolved by the Service Manager giving an instruction. The wording of the second bullet of Clause 60.1(1) means this will not qualify as a compensation event.

Also, under Clause 17.2, the Contractor notifies the Service Manager as soon as it considers that the Scope requires it to do anything which is illegal or impossible, the Parties clearly cannot contract to do something illegal. If the Service Manager agrees it gives an instruction to change the Scope appropriately.

60.1(2) The Client does not provide the right of access to the Affected Property in accordance with the Accepted Plan or the date for access shown on the latest accepted Task Order programme.

The Client has a fundamental obligation to allow the Contractor access to the Affected Property (Clause 35.1) as required. This clause recognises the dates shown in the Accepted Plan.

The Service Manager should be aware that if the Contractor shows a date in its plan when access to and use of the Affected Property or a part of the Affected Property is required and the Service Manager accepts the plan, then the Client is obliged to allow that access by that date, failing which a compensation event occurs.

60.1(3) The Client does not provide something which it is to provide by the date shown in:

- *the Accepted Plan; or*
- *the latest accepted Task Order programme.*

The Client is obliged to provide facilities and other things as stated in the Scope.

Note that, while the sub-clause refers to "the Client" it also applies to anyone who acts on behalf of the Client and who has an obligation to provide something on the Client's behalf, so this clause could include, for example, the Client not providing free issue consumables to the Contractor, or the Service Manager not providing technical information to the Contractor.

60.1(4) The Contractor receives a Task Order after the starting date stated in the Task Order.

The starting date for the Task Order is stated by the Service Manager when it instructs the Contractor to submit a quotation for the Task. If there is a delay in issuing the Task Order so that the starting date can no longer be complied with, then that is a compensation event.

60.1(5) The Service Manager gives an instruction to stop or not to start any work.

It is important to note that the Service Manager giving an instruction to stop or not to start any work is always a compensation event, regardless of the reason why the instruction was given. However, if the Service Manager gives an instruction to the Contractor to stop a part of the work, for example, because the Contractor is working unsafely and the Contractor notifies the Service Manager that it believes it is a compensation event, then under Clause 61.4, if the Service Manager decides that the event arose from a fault of the Contractor, then it notifies the Contractor that the Prices are not to be changed.

60.1(6) The Client or Others do not work in accordance with:

- *the Accepted Plan;*
- *the latest accepted Task Order programme; or*
- *the conditions stated in the Scope.*

The words "the Client or Others do not work in accordance with the Accepted Plan" can often cause debate. What work is the Client or Others required to do? Also, "work within the conditions" infers a process over a period of time rather than a specific task or date so it may be difficult to determine whether someone has worked within the time shown in the Accepted Plan?

Again, it is important that in addition to the Client showing what it is going to do, the Contractor shows these requirements and the dates for providing them in its plan submitted for acceptance, so that when the Service Manager accepts the plan it accepts liability on behalf of the Client if it fails to provide it.

Note that the term "Others" are people or organisations who are not the Client, the Service Manager, the Adjudicator, the Contractor, or any employee, Subcontractor or supplier of the Contractor. Examples of Others being utilities companies, regulatory bodies, or other perhaps specialist contractors engaged by the Client under different contracts.

Some Clients amend this clause to exclude their liability for "Others" thus passing the risk to the Contractor. This is unreasonable because the Contractor usually has no more contractual or financial control over them than the Client.

60.1(7) The Service Manager does not reply to a communication from the Contractor within the period required by this contract.

If the Service Manager does not reply to a communication, which could be for example, the Contractor submitting the name of a proposed Subcontractor or a plan, etc., and not receiving a reply, then it is a compensation event as it is fundamentally a breach of contract. However, if the Contractor notifies a compensation event in this respect, and the Service Manager decides that event has no effect upon Defined Cost then again under Clause 61.4, it notifies the Contractor that the Prices are not to be changed.

The "period required by this contract" may either be an expressly stated period of time, for example two weeks for the Service Manager to reply to the submission of a plan, or the more general "period for reply" stated in Contract Data Part 1.

60.1(8) The Service Manager changes a decision which the Service Manager had previously communicated to the Contractor.

The contract refers to "decisions" under the following clauses, underlining added for emphasis:

- Clause 50.1: The first assessment date is decided by the Service Manager
- Clause 61.4: The Service Manager decides whether an event notified by the Contractor is a compensation event
- Clause 61.5: The Service Manager decides that the Contractor did not give an early warning of the event which an experienced Contractor could have given
- Clause 61.6: The Service Manager decides that the effects of a compensation event are too uncertain to be forecast reasonably
- Clause 63.5: The Service Manager has notified the Contractor of its decision that the Contractor did not give an early warning of the event which an experienced Contractor could have given
- Clause 64.1: The Service Manager assesses a compensation event if it decides that the Contractor has not assessed the compensation event correctly

This clause has been interpreted by some NEC4 practitioners as a provision to allow the Service Manager to change and revise an assessment of an implemented compensation event. This is an incorrect interpretation, it must be stressed that the Service Manager does not have the authority to change an assessment once it has been implemented. The Service Manager's assessment, or acceptance of the Contractor's quotation, is final and the only recourse, aside from an agreement between the Client and the Contractor under Clause 12.3, would be adjudication.

60.1(9) The Service Manager withholds an acceptance (other than acceptance of a quotation for not correcting a Defect) for a reason not stated in the contract.

The clauses dealing with acceptance are:

- Clause 13.4/13.8/14.1: Acceptance of a communication
- Clause 16.2: Acceptance of the Contractor's proposal
- Clause 16.3: Acceptance of the Contractor's proposal
- Clause 21.1: Acceptance of the design of an item of Equipment
- Clause 22.1: Acceptance of the Contractor's replacement key person
- Clause 24.2: Acceptance of a proposed Subcontractor
- Clause 24.3: Acceptance of the proposed conditions of contract for a Subcontractor
- Clause 31.3: Acceptance of the Contractor's plan
- Clause 32.2: Acceptance of revised plans

- Clause 33.1: Acceptance of the Contractor's Task Order programme
- Clause 32.2: Acceptance of revised Task Order programmes
- Clause 40.2: Acceptance of the Contractor's quality policy statement and quality plan
- Clause 44.2: Acceptance of a quotation to accept a Defect
- Clause 61.4: Acceptance of a compensation event
- Clause 62.3: Acceptance of a Contractor's quotation for a compensation event
- Clause 62.6: Acceptance of a Contractor's quotation for a compensation event
- Clause 84.1: Acceptance of the Contractor's insurance certificates

Option A

- Clause 55.2 – Acceptance of a revision of the Price List

Option C and E

- Clause 26.4 – Acceptance of the proposed contract data for a subcontract

Secondary Options

- X10: Acceptance of the Information Execution Plan
- X13: Acceptance of the organisation providing the performance bond
- X19: Task Order programme
- X29: Climate Change Plan

Typically, within each of the clauses are stated the reasons for the Service Manager not accepting whatever is proposed. If the Service Manager does not accept for a different reason than that stated, then this is a compensation event.

For example, the Service Manager may decide not to accept a proposed Subcontractor as it has never heard of them. The Contractor may offer additional information about this proposed Subcontractor, but if the Service Manager still does not accept them as it has never heard of them, then the reason for non-acceptance is not that the appointment of the Subcontractor "will not allow the Contractor to Provide the Service".

60.1(10) A test or inspection done by the Service Manager causes unnecessary delay.

Under Clause 41.5, the Service Manager does its tests and inspections without causing unnecessary delay to the work, so it needs to act and carry out that test or inspection in a timely fashion. How long the test or inspection takes before it transitions from "reasonable" to "unreasonable" delay is clearly a subjective judgement, but if the Service Manager delays the Contractor beyond what would be deemed "necessary" bearing in mind the type of test or inspection, then it is a compensation event.

To avoid such delays, and thereby compensation events, the Client should set out in the Scope what tests and inspections it intends to perform, the relevant timescales, what notice will be given and what, if anything, has to be done by the Contractor.

60.1(11) A change to the Affected Property

- *by the Client or Others; or*
- *as a result of a Task Order*

The Affected Property has a significant effect on the way that the Contractor delivers the services. Where the Affected Property changes, then so may the way that the services are delivered and consequently the cost of that delivery.

Example

A Contractor maintaining hospital buildings has 25,000 lights to maintain including the emergency lighting system.

The replacement of all lights with new LED lights, paid for by the Client, will reduce the maintenance obligations as the new LED lights will fail less frequently than their predecessors. The workload for the Contractor will reduce and consequently a compensation event will occur.

The effect of the compensation event will be to amend the Prices to reflect the new (lower) workload and the likely higher cost of spare components.

60.1(12) An event which is a Client's liability stated in these conditions of contract.

The Client's liabilities are stated in Clause 80.1 but additional Client's liabilities may also be inserted into Contract Data Part 1.

The categories of Client's liability in the contract and therefore covered by this clause:

- Claims and proceedings relating to the Client's use or occupation of the Affected Property.
- A fault of the Client or any person employed by or contacted to it.
- Loss or damage to items supplied by the Client to the Contractor until the Contractor has received them, or up to the point of handover to the Contractor.
- Loss or damage to Plant and Materials owing to war, civil war and other events and circumstances.
- Loss or damage to any Equipment, Plant and Materials retained by the Client after termination.
- Loss or damage to the Affected Property and any other property owned or occupied by the Client.
- Loss or damage to any Plant or Materials after they have bene included in the Affected Property.
- Any other liabilities referred to in the Contract Data. These should be stated in Contract Data Part 1.

60.1(13) The Client does not provide materials, facilities and samples for tests and inspections as stated in the Scope.

Under Clause 41.2, the Contractor and the Client provide materials, facilities and samples for test and inspections in accordance with the Scope. If the Client does not comply in providing the facilities or is late in providing them, it is a compensation event. Clearly, if the Contractor fails, that is their risk.

60.1(14) The Service Manager notifies a correction to an assumption which the Service Manager stated about a compensation event.

Under Clause 61.6, if the Service Manager decides that the effects of a compensation event are too uncertain to be forecast reasonably, it states assumptions, the quotation being based on these assumptions (See example under Clause 61.6). If the Service Manager later notifies a correction to the assumption, it is a compensation event. In this case, the change to the Prices and any delay to Task Completion Date the Contractor had assumed in its original quotation is then replaced with corrected amounts.

Example

A Contractor cleans and empties drainage gullies and channels in the taxiway and apron areas of an airport.

Recent drainage problems have led the Service Manager to instruct additional draining and cleaning of the gullies in those areas. Prior to preparing its quotation the Contractor asked the Service Manager when these works were to take place.

The Service Manager could not reach agreement with the airport's operations department and therefore notified an assumption to the Contractor that these works would take place when the relevant areas were closed to airside traffic. Ultimately, when the gullies were emptied, they were emptied during live aircraft operations. The necessary planning and precautions led to slower progress and additional costs for the Contractor.

The Service Manager notified a correction to the assumptions to reflect the different working conditions. It was this notification that led to a compensation event.

60.1(15) A breach of contract by the Client which is not one of the other compensation events in the contract.

This is a "catch all" to cover any breaches not covered by the other compensation events, but ensures that any Client breach is assessed in accordance with the rules in the contract and avoid a potential disagreement on calculating the quantum of such a breach.

60.1(16) The Service Manager gives an instruction to correct a mistake in the Price List.

This is related to Clause 14.5, the Service Manager corrects a mistake which is:

- a departure from the method and rules stated in the Price List and used to compile it; or
- owing to an ambiguity or inconsistency.

60.1(17) The Service Manager notifies the Contractor that a quotation for a proposed instruction is not accepted or that a Task will not be instructed.

This is related to Clause 65, where the Contractor has submitted a quotation for a proposed instruction, and the Service Manager has replied to the quotation stating that the quotation is not accepted, and that the Service Manager may issue an instruction, notify the instruction as a compensation event, and instruct the Contractor to submit a quotation.

60.1(18) Additional compensation events stated in Contract Data Part 1.
This provision is discussed below.

6.4   Additional compensation events

The following are some examples of additional compensation events which may be considered:

Dealing with objects of value or of historical or other interest found within the Site

Proposed compensation event:
The Service Manager gives an instruction for dealing with an object of value or of historical or other interest found within the Affected Property.

Reason:

While a TSC will not normally include significant excavation, the possibility of finding objects of value or of historical interest, for instance while surveying properties, should be considered. Environmental hazards, such as asbestos, can also occasionally be found inside buildings.

This could include buried articles such as fossils, archaeological remains, but can also include munitions or other wartime remains. Such finds may also be governed by government legislation, local bylaws, etc. and may involve investigation by educational establishments, museums and other authorities.

In that case, it should not be considered as a Contractor's risk, as:

  (i)   it has been found on the Client's property;
 (ii)   if the Contractor sees a significant risk when it finds something such as this, it may seek to ignore it or even cover it up; and
(iii)   the Service Manager should instruct the Contractor how to deal with it.

It is therefore suggested that an additional compensation event should be considered, similar to that in the NEC4 Engineering and Construction Contract:

Searching for a Defect

Proposed compensation event:
The Service Manager instructs the Contractor to search for a Defect and no Defect is found unless the search is needed only because the Contractor gave insufficient notice of doing work obstructing a required test or inspection.
Reason:

The Service Manager may instruct the Contractor to search for a Defect. If the search shows that there is no Defect, it is a compensation event.

Note that if the Contractor did not give sufficient notice of doing work obstructing a test or inspection then it is not a compensation event.

Unforeseen physical conditions

Proposed compensation event:
The Contractor encounters physical conditions, which:

* *are within the Site;*
* *are not weather conditions; and*
* *an experienced contractor would have judged at the Contract Date to have such a small chance of occurring that it would have been unreasonable for it to have allowed for them.*

Only the difference between the physical conditions encountered and those for which it would have been reasonable to have allowed is taken into account in assessing a compensation event.
Reason:

Again, while a TSC will not normally include significant excavation, the possibility of encountering physical conditions such as natural or artificial obstructions, should be considered. This risk is possible in say highways maintenance contracts. Using the TSC in existing (possibly aged) buildings presents a risk that hazards such as asbestos might be encountered when, for example, installing new cables through service risers.

It is therefore suggested that an additional compensation event should be considered, similar to that in the NEC4 Engineering and Construction Contract:

One should also include an additional clause, again similar to the Engineering and Construction Contract that, in judging the physical conditions for a compensation event, the Contractor is assumed to have taken into account:

- the Scope: this could include site investigations, borehole data, etc.
- publicly available information referred to in the Scope – this could include reference to public records about the Service Areas
- information obtainable from a visual inspection of the Service Areas. Note the use of the term "visual", the Contractor is not assumed to have carried out an intrusive investigation of the Service Areas
- other information that an experienced Contractor could reasonably be expected to have or to obtain. This is a fairly subjective criteria, but precludes the Contractor from relying solely on what is contained within the Scope. Note the clause refers to "an experienced Contractor" not one who has a detailed local knowledge of the Service Areas

It is critical that the Client makes all information in its possession available to the Contractor. It cannot be selective, withholding information with a view of obtaining advantage.

It is also important to note that if the Contractor encounters unforeseen physical conditions, which are often, but not always ground conditions, it may not necessary be compensated for the cost and time effect of dealing with it, so it must consider carefully the wording of the contract to prove its case.

It is also important to recognise that compensation to the Contractor is assessed as the difference between what the Contractor found, and what it would have been reasonable for it to have allowed in the tender, not simply the difference between what it found and what it did allow in the tender.

Weather

Proposed compensation event:
A weather measurement is recorded:

- *within a calendar month;*
- *before the Completion Date for the whole of the services; and*
- *at the place stated in the Contract Data.*

the value of which, by comparison with the weather data, is shown to occur on average less frequently than once in ten years.

Only the difference between the weather measurement and the weather which the weather data show to occur on average less frequently than once in ten years is taken into account in assessing a compensation event.

Reason:

Unlike the NEC4 Engineering and Construction Contract, the TSC does not include a compensation event for weather-related risks.

While this may at first not seem to be an important issue for a contractor under a TSC,, it can be a concern and a major risk for someone carrying out external work which may be affected by the weather for example, external cleaning, lighting maintenance, site inspections, surveys, etc.

Example

On a TSC, a Contractor is employed to carry out a drainage survey which is estimated will take three weeks. However, in the second week, heavy rain causes the inspection to be delayed, and it is eventually completed in four weeks.

The Contractor will be liable for its own costs in any delays and possibly demobilising/ remobilising owing to the bad weather.

Where external work is involved, it may be appropriate for the Client to add a Z clause for weather related risks similar to Clause 60.1(13) of the Engineering and Construction Contract.

NB: If this additional compensation event is considered, Contract Data Part 1 would have to include for appropriate data on where weather is to be recorded, weather measurements, and weather data.

Only the difference between the weather measurement and the weather which the weather data show to occur less frequently than once in ten years is taken into account in assessing a compensation event, so the Contractor will not be compensated for the full impact of the weather event as weather likely to occur within a ten year period is the Contractor's risk, and it is assumed that the Contractor has already allowed for that in terms of price and programme.

As stated above, there are further compensation events within the following Secondary Options:

- Option X2: Changes in the law, i.e. when a change in law occurs after the Contract Date
- Option X12: Multiparty collaboration – changes to the Partnering Information, and changes to the plan caused by amendments to the timetable.
- Option X29: Climate Change – compensation events affecting the Performance Table. While X29 does not create any additional compensation events, it does require the effect of other compensation events to be considered as part of the assessment of those compensation events
- Option Y(UK)2: The Housing Grants, Construction and Regeneration Act 1996 –Contractor suspends performance owing to late/non-payment

6.5 Additional compensation events within Options A and C

Under Options A and C, the Options which use the Price List, there is an additional compensation event.

Clause 60.2

A difference between the final total quantity and the quantity stated for an item in the Price List is a compensation event if it satisfies all three bullet points within the clause.

This clause has caused some confusion, and has even led practitioners to delete it through a Z clause, as it is seen either as confusing or unfair in its application, but if we examine each of the bullet points in turn:

- *the difference does not result from a change to the Scope,*

If it had resulted from a change to the Scope, then it would have been dealt with as a compensation event under Clause 60.1(1).

- *the difference causes the Defined Cost per unit of quantity to change; and*

The purchase price of materials may be affected by a change in the quantity, for example a price reduction or increase may be applicable when the number of units exceeds or falls below a certain amount. Or the cost of mobilisation or delivery could have changed as the same vehicle delivers fewer units.

- *the rate in the Price List for the item multiplied by the total quantity which the Contractor has completed is more than 0.5 per cent of the total of the Prices at the Contract Date.*

This final bullet point compares the final extended price for that item in the Price List with the Total of the Price at the Contract Date (many contracts refer to this as the "Tender Sum" or the "Contract Sum".

If the extended price is more than 0.5 per cent of the Total of the Price at the Contract Date, and both of the previous bullet points are satisfied, then it is a compensation event. This final bullet excludes items of a 'minor value' in the Bill of Quantities.

Example

Original quantity of work

Replace faulty road gullies 150 @ £125 = £18,750

Owing to an error in measuring the road gullies when preparing the Price List, the final quantity is only 120 No.

Final quantity of work

Road Gullies 120 @ £125 = £15,000

The Total of the Prices at the Contract Date is £2,000,000

0.5% x £2,000,000 = £10,000, therefore, assuming that all three conditions within Clause 60.2 are satisfied, then it is a compensation event and a new rate/price will be established.

6.6   Notifying compensation events

Compensation events may either be notified by the Service Manager to the Contractor, or by the Contractor to the Service Manager.

Contract: Contract No:	COMPENSATION EVENT NOTIFICATION Notification No..........................
To: Contractor/Service Manager	

You are notified of the following compensation event:

<u>Description</u>

The Compensation Event is likely to have an effect on:

  ☐ **The Prices**
  ☐ **Delaying a Task Completion Date**

Has an early warning been given?

  ☐ **Yes**
  ☐ **No**

Quotation Required?

  ☐ **Yes**
  ☐ **No**

Signed: Service Manager/Contractor):............................ Date: ....

Copied to:

Contractor ☐   Service Manager ☐   File ☐   Other .... ☐

(i) The Service Manager notifies the Contractor of the compensation event at the time of giving the instruction (Clause 61.1/61.2). It also instructs the Contractor to submit quotations, unless the event arises from a fault of the Contractor or quotations have already been submitted. The Contractor puts the instruction or changed decision into effect. The Service Manager should send these notifications separately, in compliance with Clause 13.7.

(ii) The Contractor can notify a compensation event, but it must do so within 8 weeks of becoming aware of the event, otherwise it is not entitled to a change in the Prices and/or a Task Completion Date, unless the Service Manager should have notified the event to the Contractor but did not (Clause 61.3).

The intention of the clause in (ii) is to compel the Contractor to notify compensation events promptly, otherwise any entitlement to additional time and money is lost.

Note that the eight-week rule requires the Contractor to notify the compensation event to the Service Manager within eight weeks of becoming aware that the event has happened, not just to have given early warning, or mentioned the possibility of a compensation event in a discussion with the Service Manager.

Parties to TSCs are often confused about the link between early warnings and compensation events. Not all early warnings turn into compensation events and not all compensation events are preceded by early warnings.

The clause therefore covers compensation events initiated by the Contractor, rather than the Service Manager, an example being the act or omission of the Client, which the Service Manager may not be aware of unless the Contractor had notified it.

An example of the Service Manager failing to notify a compensation event when it should have would be if an instruction was issued by the Service Manager to the Contractor changing the Scope, but at the time the Service Manager did not notify the compensation event, and the Contractor in turn did not notify either. Even if the compensation event is not notified within 8 weeks, the responsibility remains with the Service Manager as it gave the instruction and should have notified the event to the Contractor but did not.

Various opinions have been published about the enforceability and effectiveness of time bars in contracts such as NEC4, commentators particularly debating whether the clause is a condition precedent to the Contractor being able to recover time and money, and whether a party, the Client can benefit from its own breach of contract to the detriment of the injured party, the Contractor.

For example, if the Client does not provide something which it is to provide by the date for providing it shown on the Accepted Plan, this is a valid compensation event under Clause 60.1(3), but can the Client prevent the Contractor from receiving any remedy because the Contractor failed to notify the Service Manager within the eight-week limit?

The authors, while not being lawyers, are of the opinion that the parties are clear as to what the terms of their contract are at the time the contract is formed, it is also clear what happens if the Contractor does not notify a compensation event within the time stated within Clause 61.3, it is protected against notifications which the Service Manager should have given but did not, and therefore the time bar must be effective and enforceable.

Contractors in particular should check the effect of Z clauses on Clause 61.3, as the eight-week limit is often reduced to a much shorter period, as low as two to three working days on one occasion seen by the authors. This subject is often brought up in adjudications and typically one party (occasionally both) cling to the arguments described earlier because of their own failure to manage the contract effectively.

A schedule of compensation events should be kept, identifying and numbering each, with additional information about each one. This is usually a default feature of online contract management systems.

Contract:				SCHEDULE OF COMPENSATION EVENTS			
Contract No:							
Ref.	Date notified	Origin? Service Manager or Contractor	Description of event	Relevant Clauses	Action by Service Manager or Contractor	Date Implemented	Notes
1							
2							
3							
4							
5							
6							
7							
8							
9							
10							
11							
12							
13							
14							

Copied to:

Contractor ☐   Service Manager ☐   File ☐   Other .... ☐

The authors have found it useful at the end of the Service Period for each Term Service Contract to review the schedule and each compensation event issued during the Service Period as "lessons learned" to consider in drafting Scopes for future TSCs.

Clause 61.4 covers three possible outcomes to the Contractor's notification of a compensation event under Clause 61.3.

Negative reply

If the Service Manager responds by stating that an event notified by the Contractor:

- arises from a fault of the Contractor;
- has not happened and is not expected to happen;
- has not been notified within the timescales set out in the contract;
- has no effect upon Defined Cost or a Task Completion Date; or
- is not one of the compensation events stated in the contract

it notifies the Contractor of its decision that the Prices and task Completion Date are not to be changed. The Service Manager only needs to name one of these as its reason for refusing a compensation event.

Positive reply

If the Service Manager decides otherwise, it notifies the Contractor that it is a compensation event, and instructs it to submit quotations.

No reply

If the Service Manager does not reply, within one week of the Contractor's notification, or a longer period to which the Contractor has agreed, then the Contractor may notify the Service Manager to that effect. If the Service Manager does not reply within two weeks of the Contractor's notification, then this failure to reply is treated as acceptance that the event is a compensation event and an instruction to submit quotations. It is a condition precedent upon the Contractor that the notification be given and two weeks pass with response before deemed acceptance can occur.

6.7 Service Manager's assumptions

When the Contractor is instructed to submit a quotation, there may be a part of the quotation which is too uncertain to be forecast reasonably. In this case the Service Manager states what the Contractor should assume (Clause 61.6).

Subsequently, when the effects are known or it is possible to forecast reasonably it notifies a correction to the assumption and it is dealt with as a correction under Clause 60.1(14) (See example above). Note, it is the Service Manager that must state the assumption, if the Contractor makes assumptions when pricing the compensation event, then they are not corrected.

6.8 Alternative quotations

Under Clause 62.1, the Service Manager may require the Contractor to provide alternative quotations, for example to carry out the service using existing resources or, alternatively, to provide additional resources which might mitigate delay. Discussion will normally take place between the Service Manager and the Contractor to ascertain what is feasible and practicable.

Once the Contractor has submitted its quotation to the Service Manager, the Service Manager is required to reply within two weeks as follows:

- it could accept the quotation, in which case the compensation event is implemented (see Clause 66.1);
- it could instruct the Contractor to submit a revised quotation, in which case it explains the reasons for doing so to the Contractor. The Contractor then has a further three weeks to re-submit it, though in most cases it will review and re-submit within a shorter time scale; or
- it could notify the Contractor that it will be making its own assessment, in which case the Service Manager has up to three weeks to make that assessment from the time of the notification.

If the Service Manager makes an assessment and the Contractor disagrees, then the Contractor can refer the matter to adjudication. There is no contractual obligation in NEC4 for the Contractor and Service Manager to "agree" the changes to the Prices, although such agreement is clearly preferable.

6.9 Quotations for compensation events

Quotations for compensation events comprise proposed changes to the Prices and also any delay to a Task Completion Date.

Contract:	Compensation Event Quotation
Contract No:	CE Quotation No...

To: Service Manager

Summary of Quotation

Change to the Prices:

Delay to Task Completion Date:

Signed: .... (Contractor)   Date: ....

The Service Manager's reply is:

☐ Notification of acceptance of the quotation
☐ An instruction to submit a revised quotation
☐ A notification that the Service Manager will be making its assessment

Signed: .... (Service Manager)   Date: ....

Copied to:

Contractor ☐   Service Manager ☐   File ☐   Other .... ☐

Submission and assessment of quotations

Contractor instructed to submit a quotation	Service Manager replies to quotation

3 weeks	2 weeks
Time for the Contractor to submit a quotation may be extended by agreement with the Service Manager (Clause 62.5)	Time for the Service Manager may be extended by agreement with the Contractor (Clause 62.5)
If the Service Manager does not agree to extend the time for submission, he notifies the Contractor that it will be making its own assessment. (Clause 62.3)	If the Service Manager does not reply within 2 weeks the Contractor may notify the Service Manager to that effect. If the Service Manager does not reply to the notification within 2 weeks, the Contractor's notification is treated as acceptance by the Service Manager. (Clause 62.6)

Quotations must be submitted within three weeks of the Contractor being instructed by the Service Manager to do so.

As stated above, the Service Manager is required to reply within two weeks of receiving the submission.

The period for submitting and replying to quotations can be extended by mutual agreement.

No reply to a quotation

If the Service Manager does not reply to a quotation within the time allowed, the Contractor may notify the Service Manager to that effect.

If the Contractor has submitted more than one quotation it states in that notification which one it proposes is to be accepted. If the Service Manager does not reply within two weeks of the notification, the notification from the Contractor is treated as acceptance of the Contractor's quotation by the Service Manager (Clause 62.6).

It is somewhat concerning that a pre-condition to deemed acceptance is that the Contractor must issue a notification giving the Service Manager a further, but final two weeks in which to deliberate, yet if the Contractor fails to provide a quotation within the time allowed, the Service Manager's notification notifies the Contractor that it will be making its own assessment, so there is no further time for the Contractor to respond.

In addition, Secondary Option W1, the Dispute Reference Table and, in W2, W2.3(8) specifically allows the Client to refer a dispute about a quotation for a compensation event which is treated as having been accepted, to be referred to the Adjudicator, so the Contractor still does not have final acceptance, although Clause 62.6 states that a failure to reply to the Contractor's notification is treated as acceptance by the Service Manager.

It is difficult to imagine how the Client might object to the factual background of such an event, but one would assume that an adjudicator whose role is to enforce the contract would have to find in favour of the Contractor provided sufficient evidence was provided? But the correct legal interpretation of these W1/W2 provisions suggests that the Service Manager's intransigence may only have a temporary downside for the Client.

6.10   Assessment of compensation events

Changes to the Prices

The starting point for assessing compensation events in the TSC is to refer to the Price List. Where the only change is to the quantity of work undertaken for an item in the Price List, then Clause 63.1 requires the Parties to amend the Prices from there.

The word "only" is important here. Where the compensation event partly includes items in the Price List and partly new items, the Price List will not be used in the assessment.

Where Clause 63.1 cannot be used, the changes to the Prices are assessed (Clause 63.2) as the effect of the compensation event upon:

- the actual Defined Cost of work done by the dividing date;
- the forecast Defined Cost of the work not yet done by the dividing date; and
- the resulting Fee.

NB: The term "dividing date" should be explained.

If the compensation event arose as a result of an instruction or changing of an earlier decision by the Service Manager, the dividing date is the date of the communication.

With all other compensation events, the dividing date is the date of notifying a compensation event.

Assessment of compensation events under Clause 63.2 is based on their forecast effect on Defined Cost. This is different from many other standard forms of contract where variations are valued using the rates and prices in the contract as a basis.

The reason for this policy within the TSC is that no compensation event is due to the fault of the Contractor or relates to a matter which is at its risk under the contract. It is therefore appropriate to reimburse the Contractor its incurred and/or forecast additional costs arising from the compensation event.

However, under Clause 63.3 if the Service Manager and Contractor agree, rates or lump sums may be used instead of Defined Cost to change the Prices. Such rates or lump sums do not have to be from the Price List (Options A and C), thereby allowing a mutually agreed fair rate or price to be agreed.

The Contractor must ensure that it includes within the quotation risk allowances for cost which have a significant chance of occurring and are its risk (Clause 63.9). This should include for say rises in the costs of materials and consumables which would not in themselves be compensation events.

Changes to the Prices take the form of changes to the Price List (Options A and C) (Clause 63.14).

In Option E contracts, the changes to the forecast amount of the Prices are notified when implementing the compensation event.

Example

A Contractor provides maintenance services to an acute hospital under a three-year Term Service Contract, using Main Option A. The services include the maintenance of car park barriers and the associated ticket vending/payment machines (TVMs).

The Client has engaged a specialist firm, H5T, to remove the barriers and the existing system and replace them with automatic number plate recognition cameras which are connected to an off-site monitoring and charging system. New payment machines have replaced the old ones. The new machines do not accept cash and only work with credit cards, smartphone apps or prepaid payment cards on sale in the hospital reception. 14 months of the *service period* remain.

The Service Manager wants the Contractor to maintain the new cameras and payment machines but is keen to avoid paying for the maintenance that is no longer required for the original machines.

The Price List contains the following entries:

Item	Unit	Rate	Price, £
101 Car park barriers – maintain	Month	160.00	5,760
102 TVMs (12 no.) – maintain	Month	42.00	18,144
103 TVMs (12 no.) – clean	Month	60.00	2,160
104 TVMs (12 no.) – empty and bank	Month	132.00	57,024
105 TVMs – consumables	At cost	At cost	At cost

The Client has received a quotation from H5T (the system's installer) to maintain all of the new site-based hardware at the rate of £3,500 per year. Provision of the back-office systems for enforcement and hardware will be contracted separately. The installer has indicated that it is willing to become a subcontractor to the Contractor but the Service Manager is unsure whether H5T should work directly for the Client or as a Subcontractor to the Contractor.

The starting point is with Clause 63.1. If the only change here is to remove items 101 to 105 from the Scope then the Price List will be used to assess the change to the Prices. In this case the total would be (for the 14 remaining months);

$$14 \times (160 + 42 + 60 + 132) = 5,516.00$$

However, if the change to the Scope included a requirement for the Contractor to provide the new maintenance services, whether subcontracted or not, then the existing items in the Price List would not be used in the assessment and an assessment of the combined effect on Defined Cost would be needed, using the four headings in Clause 63.2. That is because the new services do not yet feature on the Price List and therefore Clause 63.1 does not apply.

Delay to a Task Completion Date

Note that under Clause 62.2, within the Contractor's quotation, in addition to changes to the Prices, it also needs to consider, and where appropriate include, any delay to a Task Completion Date.

Note the clause states "any delay", not "any effect", so the Task Completion Date can only be changed to a later date or remain unchanged owing to a compensation event.

Note that, under Clause 63.6, in assessing any delay to the Task Completion Date, the delay is the length of time that due to the compensation event planned Task Completion is later than planned Task Completion as shown on the Accepted Task Completion Programme, so any terminal float is retained by the Contractor.

From these provisions one can see that no compensation event can result in an earlier Task Completion Date.

6.11   Changes to the Scope that omit work

When the Service Manager gives an instruction changing the Scope by omitting work, the method of changing the Prices is often misunderstood.

The tendency is for parties to simply delete or even to just ignore the item in the Price List, the activity in the Price List which represents the omitted work is simply reduced or not paid Admittedly, this is usually the correct way to price change in a non-NEC contract, but in an NEC contract, that assumption is not always correct.

Under Clause 63.1, any compensation event which only affects the quantities of work shown in the Price List is assessed using the Price List.

For other compensation events, for example, omitted work owing to a change to the Scope is assessed not by omitting or remeasuring the item in the Price List, but by forecasting the Defined Cost of that omitted work and adding the Fee percentage. The resulting amount is then adjusted against the value in the Price List.

This principle is clearly stated in Clause 63.2 of the contract i.e. the changes to the Prices are based on:

- the actual Defined Cost of work already done;
- the forecast Defined Cost of the work yet to be done; and
- the resulting Fee.

It must be remembered that the principle with compensation events in terms of their financial value is that neither party gains or loses as a result of the compensation event; the Contractor is compensated so that it is in the same financial position after the event as it would have been before the event.

If a Contractor has included low rates in its tender for work which is subsequently omitted it is quite probable that the application of forecast Defined Cost plus Fee will give rise to a negative value, this is testament to the fact that the Contractor would have made a loss if the work had not been omitted. Similarly if high rates exist, the Contractor will retain the margin it would have made if it had carried out the work.

Example

A highways maintenance contractor is required by the Scope to undertake a structural inspection of lighting columns and masts once a year. The industry norm, specified in an advisory note, is to make these inspections on a risk-assessment basis, typically occurring every four to six years. Recognising the savings that could be generated the Service

Manager instructed a change to the Scope to reduce the frequency of the inspections to that specified in the advisory note.

The Price List contained an entry for each inspection and therefore the total of the Prices was amended by the rate in the Price List multiplied by the reduction in inspections. This relies upon the wording in Clause 63.1.

If, instead of reducing the frequency of inspections, a change was instructed to only undertake the inspections overnight in the city centre to avoid inconvenience, then it would probably be impossible to assess the impact of the compensation event from the Price List. The Service Manager and Contractor would then need to look at the effect on Defined Cost as set out in Clause 63.2.

6.12 The Service Manager's assessments

Under Clause 64, the Service Manager must, for the following reasons, assess a compensation event:

- if the Contractor has not submitted a required quotation and details of its assessment within the time allowed;
- if the Service Manager decides that the Contractor has not assessed the compensation event correctly in a quotation and it does not instruct the Contractor to submit a revised quotation;
- if, when the Contractor submits quotations for a compensation event, the Contractor has not submitted a plan which the contract requires it to submit; or
- if, when the Contractor submits quotations for a compensation event, the Service Manager has not accepted the Contractor's latest plan for one of the reasons stated in the Contract; and
- if a Task is affected by the event and the Contractor has not submitted alterations to its Task Order programme as required by the Contract.

These are all derived from some failure of the Contractor either to submit a quotation, to assess the compensation event correctly or to submit an acceptable plan.

If the Service Manager makes its own assessment it should put itself in the position of the Contractor, giving a properly reasoned assessment of the effect of the compensation event, detailing the basis of its calculations and providing the Contractor with details of that assessment.

A Service Manager's assessment is not simply the quotation returned to the Contractor with "red pen" reductions down to a figure the Service Manager is prepared to accept.

It is also important to recognise that the Service Manager is not sending its assessment to the Contractor for acceptance, it is the final decision under the contract, the Contractor's only remedy being adjudication.

Under Clause 64.4, if the Service Manager does not assess a compensation event within the time allowed the Contractor may notify it to that effect. If the Service Manager does not reply to the notification within two weeks, the Contractor's notification is treated as acceptance of the quotation by the Service Manager. The only remedy the Service Manager has, if it later finds the quotation is not acceptable is to refer it to adjudication.

However, it must be remembered that the role of the Adjudicator and the adjudication process is to enforce the contract, therefore if the Contractor's quotation has been submitted in accordance with the contract, then the Client will probably be unsuccessful in the adjudication.

6.13   Proposed instructions

Under Clause 65, the Service Manager may instruct the Contractor to submit a quotation for a proposed instruction. Note that the Contractor has only been instructed to submit a quotation, it does not carry out the work until instructed to do so.

The Contractor submits its quotation within three weeks of being instructed to do so.

Once the Contractor has submitted its quotation to the Service Manager, the Service Manager is required to reply as follows:

- it could instruct the Contractor to submit a revised quotation, in which case it explains the reasons for doing so to the Contractor. The Contractor then has a further three weeks to re-submit it, though in most cases it will review and re-submit within a shorter time scale;
- it could issue an instruction to carry out the work and accept the quotation, in which case it is implemented;
- it could decide not to proceed with a proposed instruction; and
- it could notify the Contractor that the quotation is not accepted.

If the quotation is not accepted the Service Manager may still instruct the work to be carried out, but it is treated as a compensation event and the Contractor is instructed to submit a quotation.

6.14   Implementing compensation events

Under Clause 66, the Service Manager implements each compensation event by notifying the Contractor of the quotation which it has accepted or of its own assessment. It implements the compensation event when it accepts a quotation or completes its own assessment, or when the compensation event occurs, whichever is the latest. Where the Service Manager has not responded to the Contractor's notification on time (Clauses 62.6, 64.4) then the event is implemented when the Contractor's quotation is deemed to be accepted.

This clause emphasises the finality of the assessment of compensation events. If the subsequent records of resources on work actually carried out show that achieved Defined Cost and timing are different from the forecasts included in the accepted quotation or in the Service Manager's assessment, the assessment is not changed. The only circumstances in which a review is possible are those stated in Clause 61.6.

7 Use of Equipment, Plant and Materials

7.1 Introduction

Use of Equipment, Plant and Materials is covered within the TSC by Clauses 70.1 to 71.2.

Users of the TSC need to take care in using the terms, "Equipment" (a defined term referring to items provided by the Contractor to Provide the Service) and "equipment" (an undefined term which consequently has its ordinary meaning and relates to equipment not provided by the Contractor).

The TSC uses the term "equipment" to describe those items provided by the Client, for example in Clause 71.1.

The general principle with title to Plant and Materials brought onto a site where a service is delivered by the Contractor is that, in the absence of any contractual provision to the contrary, title will remain with the Contractor until they are incorporated (built) into the Affected Property, at which point they will become the property of the owner of the land upon which the works are being built.

Once title has passed in this way the Contractor cannot, without an express right in the contract, normally the consent of the Contract Administrator (here, the Service Manager), remove the Materials.

The TSC does not use the term "Site" but "Affected Property" and "Service Areas" which have the following meanings:

- Affected Property – the property of the Client or Others affected by the work or used by the Contractor in Providing the Service
- Service Areas – the Affected Property and those parts of the service areas which are necessary for Providing the Service, so this could include for example, a call centre operated by the Contractor to handle enquiries and call out requests

Clause 70.1 states that "*whatever title the Contractor has to Plant and Materials passes to the Client if they have been brought within the Service Areas*".

Note that in saying "*whatever title the Contractor has*", if it does not already have title itself, normally because it has not paid for it, then it cannot pass title to the Client.

It is common within term service contracts that the Contractor uses equipment, Plant and Materials provided by the Client, as the Contractor is often working within the Client's premises. If the Client gives the Contractor the right to use these facilities, then the Contractor may only use them to Provide the Service, and for no other reason.

Under Clause 70.1, if the Contractor removes Plant and Materials from the Service Areas with the Service Manager's permission, title reverts back to the Contractor.

DOI: 10.1201/9781003463771-8

Two queries immediately come to mind:

(i) One could take the requirement for permission to its extreme in that the Service Manager has to give permission for the Contractor to remove each item of redundant or waste material from the Service Areas. However, although the definition of Plant and Materials under Clause 11.2(11) refers to Materials "intended to be included within the Affected Property", waste materials were originally intended, but subsequently were never used within the Affected Property!

(ii) A Subcontractor or supplier would not be a party to the contract, so it would normally not need the Service Manager's permission to remove materials, but also note that under Clause 24.1, the contract applies as if a Subcontractor's employees and equipment were the Contractor's.

Note that NEC4 has no express right for the Contractor to be paid for unfixed materials either within or outside the Service Areas (normally referred to in other contracts as "materials on site" or "materials off site").

One has to refer to the specific Main Option clauses to determine any entitlement, for example under Options C and E, payment to the Contractor is based on Defined Cost paid by the Contractor before the next assessment date plus the Fee. In respect of the Defined Cost of Materials, where those Materials are being stored is not relevant to payment entitlement.

Note under Clause 71.2, the Contractor must, at the end of the Service Period:

- return to the Client any equipment, and surplus Plant and Materials provided by the Client;
- provide items of Equipment for the Client's use as stated in the Scope; and
- provide information and other things as stated in the Scope.

7.2   Payment for Plant or Materials located outside the Service Areas

When writing the Scope and/or Z clauses when making provision for payment for unfixed Plant or Materials outside the Service Areas, one should also consider including the following criteria:

- The Plant and Materials must be intended to be incorporated into the Affected Property.
- The Plant and Materials are in accordance with the contract.
- Nothing remains to be done to the Plant and Materials prior to being incorporated into the Affected Property, for example a steel staircase should look like a steel staircase, not a pile of unidentifiable stock steel.
- The Plant and Materials have been set aside from other stock, and labelled as destined for the specific service to which the contract relates.
- The Plant and Materials are vested in the Contractor, which then transfers title in accordance with the contract, free from any charges or encumbrances.
- The Plant and Materials are insured against loss or damage for their full value under a policy of insurance protecting the interests of the Client and the Contractor until they are delivered to the Service Areas.
- The Plant and Materials can be inspected at any time.
- The Plant and Materials cannot be removed from the storage location, except for delivery to the Service Areas.

The issue of title to Plant and Materials does not tend to cause many problems unless a party becomes insolvent, when it becomes a major problem as a party such as a Client who believed it had good title to unfixed Plant and Materials because the contract stated that when they were within the Service Areas it had title, finds that it does not!

While this book is intended for international use rather than purely focusing on English law, it is not unknown for a client to pay a Contractor for unfixed materials on site, but owing to the insolvency of a Subcontractor and the Contractor not having title to those materials; the Client then also had to pay the Subcontractor!

One aspect to bear in mind in considering the words "whatever title the Contractor has to Plant and Materials" is the use of "retention of title" clauses.

A retention of title clause, which normally states *"title of the goods remains with the seller until the purchase price is paid in full"* is one that allows the supplier of materials to retain ownership of those materials until they are fully and finally paid for irrespective of any contractual provisions or agreements between say, the Contractor and the Client. Sometimes a retention of title clause may state *"title in the goods remains with the seller until the price _and all other sums owing by the purchaser to the seller_ are paid in full"*.

Such clauses will override any clauses in the main contract that may stipulate that title will pass to the Client when materials are delivered to the site, when they are marked, or when the Client has paid for them as the supplier is not party to the contract. These clauses then provide security for the supplier against the purchaser's insolvency.

In order to be effective, retention of title clauses must be stipulated and accepted at the time the contract is formed i.e. when the supplier's quotation is offered and accepted.

In order to provide this protection when insolvency occurs, the receiver or liquidator must be satisfied on each of the following:

* that the wording of the clause includes the specific goods for which protection is being claimed;
* that the clause has been incorporated into the contract for the supply of those materials;
* that any goods claimed can conclusively be identified; and
* that the goods supplied relate to an unpaid invoice.

The goods being claimed under the retention of title clause must be identifiable and the receiver must be fully satisfied that any items claimed are the actual goods supplied by the supplier claiming them. Most retention of title clauses relate to raw materials, stock in trade or livestock.

The best method of identification is where the goods are marked with the name of the supplier and the name of the Client, or where their serial numbers are quoted on any unpaid invoices. The supplier should be allowed access to the insolvent's premises to inspect the goods which it considers are subject to its claim. Of course, the supplier should not be allowed to remove any goods until the official receiver is satisfied that the claim is valid.

If a retention of title clause was drafted only to retain ownership of the goods until payment was made for them, goods such as any raw materials cease to be caught by it once the manufacturing process has begun, i.e. once the goods have lost their identity.

For example, leather supplied in the production of safety boots, once cut is deemed to have created a new product.

It may be possible for the supplier to retain title to the goods supplied even if they have been used in a manufacturing process provided they are still identifiable, in their original form and are easily removable.

For example, ceiling tiles supplied to be installed into a suspended ceiling do not lose their identity because they are fixed within the ceiling grid. However, they must be capable of being removed without damaging the ceiling grid.

8   Liabilities and insurance

8.1   Introduction

Liability and insurance are covered within the TSC by Clauses 80.1 to 86.3.

Insurance is the principle of many paying in for the few to be compensated. It is basically a contract between the "insured" who offers to pay an agreed sum, a premium, to the "insurer" who warrants to pay out an agreed sum should a particular and specified event occur e.g. damage to property, death or bodily injury, etc.

In effect, it is possible to insure against anything, which holds an element of risk.

The required insurances under a service contract can be grouped under the following headings:

(i)   Loss or damage to Plant, Materials and Equipment

This covers loss or damage to Equipment and any unfixed Plant and Materials on site.

(ii)   Public Liability

This is to cover the insured against any death, injury or damage claims from members of the public, other than their own employees.

(iii)   Employer's Liability

In many countries, employers of staff have a statutory obligation to provide cover for their employees in the event of death, injury or damage caused during the course of their employment.

8.2   Liabilities and insurance provisions within the TSC

Within the TSC, Clause 80.1 lists the Client's liabilities. However, these are the Client's liabilities only in terms of loss or damage, and are not an exhaustive list of all the liabilities the Client bears under the TSC. There are many other risks such as actions or inactions of the Service Manager which are covered elsewhere as compensation events.

The categories of Client's risk within this clause are:

(i) Risks relating to the Client's use or occupation of the Affected Property, negligence, breach of statutory duty or interference with any legal right, or a fault in design.

(ii) A fault by the Client or someone employed or contracted to it.

DOI: 10.1201/9781003463771-9

(iii) Loss of, or damage to, items supplied by the Client to the Contractor until the Contractor has received them, or up to the point of handover to the Contractor. The Client should ensure that it has adequate insurance in this respect, or again ensure that Others who supply items have such insurance.

(iv) Risks relating to loss or damage to the works, Plant and Materials, caused by matters outside the control of the Parties e.g. war, civil war, strikes, riots and civil commotion not confined to the Contractor's employees, or radioactive contamination.

(v) Loss or damage to any Equipment, Plant and Materials retained by the Client after a termination except owing to activities of the Contractor.

(vi) Loss of or damage to the Affected Property and any other property owned or occupied by the Client, unless it is in connection with the Contractor Providing the Service.

(vii) Loss or damage to and Plant and Materials after they have been included in the Affected Property.

(viii) Any additional Client's liabilities referred to in the Contract Data. These should be stated in Contract Data Part 1.

The Contractor's liabilities are listed in Clause 81.1.
These include:

(i) Claims and proceedings from Others in connection with the service provision.

(ii) Loss of or damage to Plant and Materials prior to their incorporation into the Affected Property, and also loss or damage to Equipment provided by the Contractor, and to equipment provided by the Client (note the distinction between "Equipment" and "equipment").

(iii) Loss of or damage to the Affected Property and any other property owned or occupied by the Client caused by the service delivery.

(iv) Death or bodily injury to the Contractor's employees.

Each party agrees to pay the other for any claims or proceedings which are at its liability, although if a party partly contributed to the event, they will be partly liable.

The Insurance Table in the contract itemises the insurances that the Contractor has to effect together with the minimum amount of cover or minimum limit of indemnity.

The default is that the Contractor provides, maintains and pays for the insurances, unless the Client states otherwise in Contract Data Part 1.

Whether the Contractor or the Client takes out the insurances, the first two insurances are always effected as a joint names policy effective from the starting date until the end of the Service Period or until a termination certificate has been issued.

The effect of Joint Names Insurance is that each party has its own rights under the policy and can therefore claim against the insurer. Each insured should comply with the duties of disclosure and notification.

The three insurances listed within the Insurance Table are:

(i) Loss or damage to Plant and Materials and Equipment.

The Contractor's All Risks (CAR) policy should cover this. The cover must include replacement costs, plus for the amounts stated in the Contract Data for the replacement of any Plant and Materials provided by the Client. The reference to "replacement cost" means the cost of replacement with Equipment of similar age and condition rather than "new for old".

(ii) Loss or damage to property (except Plant and Materials and Equipment) and liability for bodily injury or death of someone (not an employee of the Contractor) arising from or in connection with the Contractor Providing the Service.

This requires the Contractor to indemnify the Client against any loss, expense, claim, etc. in respect of any personal injury or death caused by the carrying out of the work, other than their own employees. This includes the liability toward members of the public who may be affected by the construction work, although they have no part in it. In the case where a party makes a claim directly against the Client owing to a death or injury the Contractor should either take on that claim, or alternatively the Client can sue the Contractor to recover any monies.

The Contract Data states the minimum amount of cover or minimum limit of indemnity with cross liability so that the insurance applied to both Parties separately. The Contractor could be liable for whatever the amount of any claim, therefore it must consider the minimum value in the Insurance Table purely as a guide.

The Contractor must be able to prove, if requested by the Client, that it has the required cover in place, as if it does not, the Client may take out insurances himself and deduct the cost of the premiums from the Contractor.

(iii) Death of, or bodily injury to, employees of the Contractor arising out of and in the course of their employment in connection with the contract.

This covers the liabilities of the Contractor as an employer of people to insure against injury or death caused to people while carrying out their work, which is a legal obligation in most countries. The Contractor must be able to prove, if requested by the Client, that it has the required cover and the premiums are up to date, as if it does not, the Client may take out Insurances itself and charge the cost of the premiums from the Contractor.

The minimum amount of cover is the greater of the requirements of the applicable law and the amount stated in the Contract Data for any one event.

8.3   What is Contractors' All Risks Insurance?

Contractor's All Risks Insurance normally covers damage to property, such as damage to buildings, infrastructure or other structures being worked on. It also covers liability for third party claims for injury and death or damage to third party property.

This insurance is usually taken out in the joint names of the Contractor and the Client. Other interested parties, such as funders, often ask to be added as a joint name. The theory is that if damage occurs to the insured property then, regardless of fault, insurance funds will be available to allow for reinstatement.

8.4   Insurance of neighbouring property

Some contracts have provision for insuring neighbouring property which may suffer from collapse, subsidence, vibration, etc. owing to damage by, for example piling, deep excavations or demolition works, where not caused by the negligence of the Contractor.

The TSC does not provide for this, and in effect the Contractor under a TSC would be unlikely to be carrying out this type of work anyway, but it can be added as a Z clause if applicable.

The Client should consider its liabilities in respect of neighbouring owners who may be affected by the service and if necessary either instruct the Contractor to provide such insurance, or take it out themselves.

Such a policy would normally exclude:

(i) Damage caused by the negligence of the Contractor or its Subcontractors.
(ii) Damage which can be reasonably foreseen e.g. cracking may have already occurred to a neighbouring building prior to the works commencing. In this respect, it is important that a condition survey with photographic evidence be carried out before construction commences.
(iii) Where the damage is caused by any excepted risks identified in the contract. These risks will normally be held by the Client.

8.5 Professional Indemnity (PI) or Professional Liability Insurance

Contract Data Part 1 provides for the Client to insert a requirement for the Contractor to provide additional resources.

Where the Contractor is designing parts of the works, for example under a Task Order, it would be advisable for the Client to include a requirement for Professional Indemnity Insurance.

If a Contractor or Consultant providing a service to a Client makes a mistake, is found to be negligent, or gives inaccurate advice, then it will be liable to the Client in event that the Client incurs a loss as a result. This loss can be very significant where the design has to be corrected, parts of the structure have to be taken down and reinstated, a facility has to be closed down while the remedial measures take place, and there are also legal costs.

Professional Indemnity claims can arise where there is negligence, misrepresentation or inaccurate advice which does not give rise to bodily injury, property damage or personal injury, but does give rise to some financial loss.

Additional coverage for breach of warranty, intellectual property, personal injury, security and cost of contract can be added.

In that event, although the Client claims against the Contractor or Consultant rather than from the insurers, Professional Indemnity Insurance protects the Contractor or Consultant against claims for loss or damage made by a client or third party.

Note that Professional Indemnity Insurance policies are based on a "claims made" basis, meaning that the policy only covers claims made during the policy period when the policy is "live", so claims which may relate to events occurring before the coverage was active may not be covered.

However, these policies may have a retroactive date which can operate to provide cover for claims made during the policy period but which relate to an incident after the retroactive date. The Service Manager, on behalf of the Client should ensure that the Contractor has taken out and maintained the required insurances for the full period of its liability.

Claims which may relate to incidents occurring before the policy was active may not be covered, although a policy may sometimes have a retroactive date, earlier than when the policy was created.

8.6 Proof of insurance

The Contractor is required to submit to the Service Manager for acceptance certificates of insurance as required by the contract signed by the insurer or the insurance broker:

(i) before the starting date
(ii) on each renewal of the insurance policy

If the Contractor does not insure or submit a required certificate, the Client may insure a risk, the cost of the premium to the Client is then paid to the Client by the Contractor.

If the Client is to provide insurance, then the Service Manager submits policies and certificates to the Contractor for acceptance before the starting date and as and when instructed by the Contractor.

If the Client does not submit a required certificate, the Contractor may insure a risk, the cost of the premium being paid by the Client.

9 Termination

9.1 Introduction

Termination is covered within the TSC by Clauses 90.1 to 93.2, and also by Clauses 93.3 and 93.4 under Option C There are also additional references to termination in Secondary Options X11 (Termination by the Client) and X19 (Termination by either Party) which are discussed in another Chapter.

Termination is a word most commonly used in the context of construction contracts to refer to the ending of the Contractor's employment under the Contract.

In the UK, legal jurisdictions, and several others, the parties have a common law right to bring the contract to an end in certain circumstances, but most standard forms give the parties additional and express rights to terminate upon the happening of specified events and for the processes that will apply thereafter.

Some contracts refer to termination of the contract, while others refer to the termination of the Contractor's employment under the contract. In practice, it makes little difference, because most contracts make express provision for what is to happen after termination, with the parties retaining certain rights and obligations.

It should be noted that most contracts require a notice to be given by one party to the other, unless the breach is due to insolvency the offending party then has a period of time in which to remedy the breach, failing which termination can take place.

It is important to note that neither party should attempt to terminate employment under the contract unless they are sure that the provision is available within the contract. If termination is held to be wrongful it is usually a repudiation of the contract.

It must be stressed that termination is an entitlement not an obligation. Clearly if a Party becomes insolvent there is no choice, but the Parties should always consider other alternatives if possible.

9.2 Termination provisions within the TSC

Within the TSC, there are 21 reasons listed for either Party to terminate:

- **Reasons R1 to R10** provide for either Party to terminate due to the insolvency of the other. This includes bankruptcy, appointment of receivers, winding-up orders and administration orders dependent on whether the Party is an individual, a company or a partnership.
- **Reasons R11 to R13** provide for the Client to terminate if the Service Manager has notified the Contractor that the Contractor has defaulted and has not remedied the default within four weeks of the notification of default in respect of the Contractor substantially failing to

DOI: 10.1201/9781003463771-10

comply with its obligations, not providing a bond or guarantee, or appointing a Subcontractor for substantial work before the Service Manager has accepted the Subcontractor.

It is quite subjective as to what the words "substantially failed to comply" and "substantial work" means within these provisions. One would assume that it is not intended to relate to minor breaches, but the Contractor could "substantially fail to comply" with a minor obligation?

- **Reasons R14 to R15** provide for the Client to terminate if the Service Manager has notified that the Contractor has defaulted and the Contractor has not stopped defaulting within four weeks of the notification in respect of substantially hindering the Client or Others or substantially broken a health or safety regulation and not stopped defaulting within the four weeks of the notification.

   Again, the use of the word "substantially" is quite subjective. How would one substantially hinder someone, or substantially break a health or safety regulation? Isn't a health or safety regulation either "broken" or "not broken"?

- **Reason 16** provides for the Contractor to terminate if the Client has not paid an amount certified by the Service Manager within thirteen weeks of when it should have been paid.

- **Reason 17** provides for either Party to terminate if the Parties have been released under the law from further performance of the whole of the contract. Note the reference to "the whole of the contract", not just a part.

- **Reasons 18 to 20** provide for the Service Manager having instructed the Contractor to stop or not restart any substantial work or all work and an instruction allowing the work to restart or start has not been given within 13 weeks.

   Either Party may terminate if the instruction was due to a default by the other, or if the instruction was due to any other reason.

- **Reason R21** provides for the Client to terminate if the Contractor does a Corrupt Act, unless it was done by a Subcontractor or supplier, and the Contractor was not and should not have been aware of the Corrupt Act, or the Contractor informed the Service Manager of the Corrupt Act and took action to stop it as soon as it became aware of it.

   Note that a Corrupt Act is defined under Clause 11.2(4) as "*the offering, promising, giving, accepting or soliciting of advantage as an inducement for an action which is illegal, unethical or a breach of trust, or abusing any entrusted power for private gain. . . . .in connection with this contract, or any other contract with the Client*. This includes any commission paid as an inducement which was not declared to the Client before the Contract Date"

Note that, in addition, under Secondary Option X11, the Client can terminate for a reason not stated in the Termination Table.

While this may seem inequitable, if the Client terminates for a reason not stated in the Termination Table the Client will have to pay the Contractor for services provided up to the termination, other costs incurred in the expectation of completing the whole of the service for example orders placed and other commitments made, the forecast Defined Cost of removing Equipment, and dependent on the Main Option used, the direct fee percentage applied to the difference between the original total of the Prices (at the Contract Date) and the Price for Service Provided to Date, essentially a loss of profit/overheads provision. Clearly the earlier the Client exercises this right of termination, the greater this amount.

Also, under Secondary Option X19, after the minimum period of service stated in the Contract Data a Party may terminate the Contractor's obligation to Provide the Service, for a reason not stated in the Termination Table.

If they do so, they notify the Service Manager and the other Party. The Service Manager then issues a termination certificate at the end of the notice period following the notification.

If a Party terminates after the minimum period of service the Client may complete the service and use any Plant and Materials provided by the Contractor, and the Contractor provides to the Client information and other things which the Scope says ere to be provided at the end of the Service Period. In addition, the Contractor leaves the Service Area and removes its Equipment.

The Contractor is paid (Amount A1):

- an amount due assessed as for normal payments
- the Defined Cost for Plant and Materials which have been delivered and retained by the Client
- other Defined Cost in reasonable expectation of completing the whole of the service (such as long-term supply contracts for consumables)
- any amounts retained by the Client.

With all the termination reasons, the Party wishing to terminate notifies the Service Manager and the other Party giving details of its reasons for terminating.

The Service Manager then issues a termination certificate to both Parties if it is satisfied that the reason for termination is valid under the contract.

It is perhaps curious that the Service Manager, who acts for the Client makes the decision as to whether the termination is valid?

Procedures on Termination

Under any of the reasons, whoever terminates, and whether the reason is included in the Termination Table or not, the Client may complete the service, either themselves or using another Contractor, and may use any Plant and Materials provided by the Contractor. This is Procedure P1.

Also under any of the reasons, whoever terminates, and whether the reason is included in the Termination Table or not, the Contractor must provide to the Client information and other things that the Scope says it is to provide at the end of the Service Period. For example, this might include data for asset management systems and helpdesk systems or equipment provided free-issue by the Client. This is Procedure P4.

Client Terminates

For reasons R1 to R15, R18 or R21 and for a reason other than that stated in the Termination Table, the Client may instruct the Contractor to remove any Equipment, Plant or Materials which belong to the Contractor and assign the benefit of any subcontract to the Client. This is Procedure P2.

Also, for reasons R1 to R15 or R18 or R21, the Client may use any Equipment to which the Contractor has title and complete the service. The Contractor "promptly" removes such Equipment when instructed to do so. Clearly in this respect, the Contractor does not have title to hired Equipment; also if the Equipment is owned by the Contractor but it has become insolvent, then some discussion may have to take place between the Client and the relevant insolvency practitioner or other authorised representatives. This is Procedure P3.

For reasons R17 or R20, the Contractor leaves the Service Areas and removes the Equipment. This is Procedure P4.

Contractor Terminates

Whatever the reason, the Contractor follows procedures P1 and P4 are described above.

Payment on Termination

Under any of the reasons whether included in the Termination Table or not, on termination, the Contractor is entitled to be paid:

Amount A1

- an amount due as for normal payments
- the Defined Cost for Plant and Materials which have been delivered and retained by the Client or which the Client owns and which the Contractor has to accept delivery
- other Defined Cost in reasonable expectation of completing the whole of the service (such as long-term supply contracts for consumables)
- any amounts retained by the Client

Amount A2

If the Client terminates for Reasons R17 or R20 or the Contractor terminates for any reason the Contractor is paid the forecast cost of removing Equipment.

Amount A3

If the Client terminates for Reasons R1 to R15, R18 or R21, a deduction is made of the forecast of the additional cost to the Client of completing the whole of the service.

Amount A4

If the Client terminates for a reason other than that stated in the Termination Table as stated above, dependent on the Main Option used, the direct fee percentage applied to the difference between the original total of the Prices and the Price for Services Provided to Date. If the Contractor terminates for reason R1 to R10, R16 or R19, the same provision applies.

Note that under Clauses 93.3 and 93.4 (Option C only) if there is a termination, the Contractor's share is assessed by the Service Manager following the issue of the termination certificate.

The Price for Service Provided to Date is the Defined Cost which the Contractor has paid and is committed to pay, and the total of the Prices (the target) is the quantity of work which the Contractor has completed for each item in the Price List, and a proportion of each lump sum which the Contractor has completed.

9.3 Termination and Subcontractors

The Parties to a TSC should also consider how termination of their contract affects Subcontractors. If the Contractor is the defaulter, termination occurs and the Client wishes to proceed with the works, any subcontracts would be assigned to the Client.

However, if the Contractor is the terminating party it should ensure that provisions are in place regarding termination of subcontracts, and termination provisions in the main contract should be mirrored in the subcontracts.

10 Resolving and avoiding disputes

10.1 Introduction

Resolving and avoiding disputes is covered within the TSC by Options W1 and W2.

Options W1 and W2, cover the adjudication process, whether the British-based legislation, the Housing Grants, Construction and Regeneration Act 1996 is applicable (Option W2), or not applicable (Option W1). The Act applies in England, Wales and Scotland, with the Construction Contracts (Northern Ireland) Order 1997 applicable in Northern Ireland.

This legislation applies to work being undertaken that the Act or Order describe as construction operations. A detailed explanation falls outside this book but many service contract works require the contract under which they are procured to comply with certain dispute resolution provisions.

Note, the Housing Grants, Construction and Regeneration Act 1996 was supplemented by the Local Democracy, Economic Development and Construction Act 2009.

If the TSC is used within the UK, the Client must, by reference to the Act, determine whether the contract is a "construction contract", and select the appropriate dispute resolution Option. If the TSC is used outside the UK, Option W1 will always apply.

Clearly, there is nothing to prevent the parties jointly attempting to resolve a dispute as they see fit, by a consensual and non-binding route such as negotiation, mediation, conciliation or expert determination, prior to, or instead of referring it to adjudication.

10.2 What is adjudication?

Adjudication has been defined as *"a summary non-judicial dispute resolution procedure that leads to a decision by an independent person that is temporarily binding but which may subsequently be reviewed by means of arbitration, litigation or by agreement"*.

It is therefore an intermediate dispute resolution process which is binding upon the parties unless and until the matter is referred to a tribunal, normally arbitration or litigation.

In the UK, adjudication is a statutory right, but it has always been included as the primary means to resolve a dispute within the NEC contracts, failing which the parties may refer a dispute to a tribunal. In fact, the majority of adjudication decisions are accepted by parties as the final result and are never taken to the tribunal.

10.3 Who can be an Adjudicator?

While many professional bodies hold registers of Adjudicators there is no formal qualification to become an Adjudicator, though there are commonly acceptable requirements which include:

DOI: 10.1201/9781003463771-11

- A recognised professional qualification in a relevant discipline. The relevant discipline then relates to the dispute in question, for example if the dispute is related to services installations, a Mechanical or Electrical Engineer would probably be a relevant discipline, though it must be remembered that the Adjudicator's role is to enforce the contract, so in that case the adjudicator does not necessarily have to be a Mechanical or Electrical Engineer just because the dispute is related to services. Clearly, as disputes are always of a legal nature, lawyers tend to make up a good proportion of the adjudicating community.
- Substantial number of years' experience within the industry, normally ten to 15 years post-qualification in a relevant discipline.
- A qualification, or at least a thorough understanding of contracts and legal principles, though in many cases the Adjudicator is not a lawyer, nor has to be.
- Recognised training in the law and practice of adjudication.
- Membership of a professional institution.
- Some also say that the Adjudicator should have Professional Indemnity insurance, though it must be emphasised that as the Adjudicator is not liable to the Parties for any action or inaction on its part other than in bad faith, it could be questioned as to whether such insurance should be a prerequisite to appointment. Most adjudicator nominating bodies require their panel members to carry appropriate insurance.

As the use of adjudication in the resolution of contractual disputes has grown, particularly in the UK where parties have a statutory right to refer a dispute to adjudication, many education establishments and professional bodies now provide diploma and certificate courses in adjudication.

10.4   Who pays for the adjudication?

The cost of the Adjudicator's fee is allocated by the Adjudicator between the parties, the Adjudicator deciding what the proportion is to be paid by each party, with in some cases one party paying the majority or even all of the Adjudicator's fee. However, both parties are jointly and severally liable for payment of the Adjudicator's fee, therefore if one party will not or cannot pay its share, possibly owing to insolvency, the other party must pay it, and recover the amount directly from the defaulting party.

Each party bears its own costs which is different to, for example, litigation where both parties' costs are usually borne by the losing party.

10.5   What will the Adjudicator charge?

There are no set hourly rates for Adjudicators, each Adjudicator will advise the party of their hourly rate and whether there are other costs payable, for example for travel, overnight accommodation or expert costs. The Adjudicator is required to properly record the time spent carrying out its duties.

10.6   What if a party does not comply with the Adjudicator's decision?

If a party does not comply with the Adjudicator's decision, the other party can refer to the courts, e.g. the Technology and Construction Court in England and Wales for enforcement of the Adjudicator's decision.

10.7   Selecting the Adjudicator

Within Contract Data Part 1 there are three options for selecting the Adjudicator:

(i)   The name and address of the Adjudicator may be stated by the Client in Contract Data Part 1 and they are then appointed before the starting date.

This has the advantage that tendering Contractors are aware at the time of tender who will be the Adjudicator in the event that a dispute arises and, if necessary, they can address the issue and object to the choice of Adjudicator within their tender submissions. Also, the Adjudicator is already in place should a dispute arise.

However, the named Adjudicator may require some form of retainer fee for being named in the contract and being available in the event that a dispute arises.

(ii)   The parties can mutually agree to the name of the Adjudicator in the event that a dispute arises. This has the advantage that there is no "Adjudicator in waiting" and therefore no fee payable.

Many TSC practitioners say that pre-appointing the Adjudicator as in option (i) is resigning oneself to the fact that there will at some point be a dispute. However, once parties are in dispute, they then have to agree who is to be the Adjudicator and appoint them, though, at this point they may not wish to agree with each other about anything!

(iii)   The name of the Adjudicator nominating body may be stated by the Client in Contract Data Part 1.

This is probably the favoured option as an independent name can be put forward by the nominating body, which is well prepared to offer the name of an Adjudicator promptly from its registers and have them appointed within the timescales set by the contract and if necessary, the appropriate legislation.

In all cases, the Adjudicator must be impartial, i.e. the Adjudicator should not show any bias towards either party. In addition, all correspondence from the Adjudicator must be circulated to both parties.

Any request for an Adjudicator must be accompanied by a copy of the Notice of Adjudication issued by the referring party, and the appointment of the Adjudicator must take place within seven days of the submission of the Notice of Adjudication to the other party.

10.8   Adjudicator Nominating Bodies

Adjudicator Nominating Bodies, as the name suggests, are organisations that fulfil the role of nominating Adjudicators. These bodies keep registers of Adjudicators with appropriate expertise and based at various geographical locations, who can act for parties should they be nominated.

Clearly, in naming the Adjudicator Nominating Body in the contract it is advisable to name an organisation with expertise in the service to be carried out.

These bodies are very knowledgeable about appointment of Adjudicators and the relevant timescales, and for a modest fee can nominate a suitably qualified Adjudicator to suit the parties' requirements.

10.9 Preparing for adjudication

Once the adjudication process starts by the issuing of a Notice of Adjudication by the referring party it is a very quick procedure, therefore it is critical that the referring party, and in turn the responding party, prepare well in the short time they have available. Time scales are stated within the contract, though these may be extended by agreement between the parties.

As the process does not start until the referring party initiates the process by issuing the Notice of Adjudication to the other party, the referrer does have the advantage of time, and in many cases surprise, as it can prepare the case before issuing the notification

Before considering initiating the adjudication process, parties should consider the following:

(i) There must be a dispute – while this may seem fairly obvious it is essential that a dispute actually exists between the parties in that one party has made a claim, which the other party denies (probably repeatedly), and that the parties have exhausted the means to resolve the dispute themselves for example by the use of negotiation, mediation, conciliation, etc., though they do not have to prove that they have tried other methods.

(ii) The dispute arises under or in connection with the contract – the Adjudicator is appointed to decide a dispute under the contract; therefore it is essential that the dispute arises on a matter covered by the contract.

(iii) The role of the Adjudicator is to enforce the contract, not to decide who is the more deserving party. It is therefore essential that, before commencing adjudication, that the referring party ensures that it has properly complied with the contract, otherwise it may find himself in a weak position.

There have been adjudications in the past where a Contractor has acted in good faith but not strictly in accordance with the contract, accepting and willingly acting on verbal instructions in order not to damage what appeared to be a good relationship with the Client, then when a dispute arises the Adjudicator states that as the contract did not provide for verbal instructions, the Contractor was incorrect to act on them, and therefore the Contractor loses the adjudication.

(iv) While many people say that, as adjudication is a fairly simple process, it should not include lawyers as the issues are industry-related rather than legal issues, it is important that the parties carry out their obligations within the short time scale they have, so if a party is uncertain as to what they have to do, and when and how they have to do it, they obtain professional advice. Adjudication under NEC contracts is a condition precedent to any further legal action.

(v) It is essential that a clear and concise Notice of Adjudication is prepared by the referring party as it defines what is in dispute, and the matters that are being referred to the Adjudicator and what they are seeking form the Adjudicator. The Adjudicator cannot consider issues which are not set out within the Notice of Adjudication.

The notice should include:

- Brief details of the service being carried out under the contract.
  While the Adjudicator may have been named in the contract, they may have little knowledge about the actual service being provided.

- The names and addresses of the parties to the contract, and therefore the parties in dispute.
- The nature and a brief description of the dispute.
- The nature of the redress which is sought from the Adjudicator.

It is critical that the responding party, and also the Adjudicator, clearly understand what redress the referring party is seeking by referring the matter to adjudication.

(i) Details included within the referral should be clearly and concisely presented so that the responding party knows what it needs to respond to, and also the Adjudicator clearly understands what is required of them. Any supporting information should be clearly referenced and relative to the matter in dispute.

    Neither party should not include or rely on any information or evidence which the other party has not seen. If they do, the other party may raise an objection, which could bring the adjudication process to a premature end without a resolution.

(ii) The contract provides for time scales to be extended by agreement between the parties, so if more time is needed in order to prepare and submit documentation, the other party should be consulted.

10.10 Statutory right to adjudication (British-based contracts)

The Housing Grants Construction and Regeneration Act 1996 applies to construction contracts and apart from a few notable exceptions, it provides a statutory right to refer a dispute under a construction contract to adjudication.

    This Act has now been amended by the Local Democracy, Economic Development and Construction Act 2009.

    The Act applies to all what it terms as "construction contracts". That term captures many activities that feature in service contracts even though construction in its normal definition might not be involved, including:

- all normal building and civil engineering work, including construction, alteration, repair, maintenance, extension and demolition or dismantling of structures forming part of the land and works forming part of the land, whether they are permanent or not;
- the installation of mechanical, electrical and heating works and maintenance of such works;
- cleaning carried out in the course of construction, alteration, repair, extension, painting and decorating and preparatory works;
- agreements with consultants such as architects, other designers, engineers and surveyors;
- elements such as scaffolding, site clearance, painting and decorating;
- labour only contracts; or
- contracts of any value.

It excludes:

- work on process plant and on its supporting or access steelwork on sites where the primary activity is nuclear processing, power generation, water or effluent treatment, handling of chemicals, pharmaceuticals, oil, gas, steel or food and drink;
- supply only contracts, that is contracts for the manufacture and/or delivery to site of goods and materials, where the contract does not provide for their installation;

- extracting natural gas, oil and minerals;
- purely artistic work;
- off-site manufacture;
- contracts with residential occupiers. This means a contract where work is carried out on a dwelling which one of the parties occupies or intends to occupy as its residence, rather than simply a housing project;
- PFI contracts. This only includes the head agreements in PFI projects, not a contract for the construction of the works; and
- finance agreements. This includes lending agreements with banks and other funders.

The Adjudicator is appointed jointly by both parties who are jointly and severally liable for those fees.

Under the Act, the contract must include adjudication provisions which comply with the requirements of the Act. If the contractual clause does not comply then the "Scheme for Construction Contracts" will apply instead. Option W2 in the NEC4 TSC complies with the Act, so the Scheme will not usually be relevant.

The provisions in the contract must:

- enable the parties to give notice at any time;
- provide a timetable with the object of securing the appointment of the adjudicator and referral of the dispute to them within seven working days;
- require the Adjudicator to reach a decision within 28 days of referral or such longer period as is agreed by the parties after the dispute has been referred;
- allow the Adjudicator to extend the period of 28 days by up to 14 days with the consent of the party by whom the dispute was referred;
- impose a duty on the Adjudicator to be impartial;
- enable the Adjudicator to take the initiative in ascertaining the facts and the law;
- provide that the decision of the Adjudicator is binding until determined by litigation, arbitration or by agreement; and
- provide that the Adjudicator is not liable for anything done or omitted in the discharge or purported discharge of its functions as Adjudicator unless the act or omission is in bad faith, and that any employee or agent of the Adjudicator is similarly protected from liability.

10.11 Does a "dispute" exist?

Only a dispute between the contracting parties can be referred to adjudication. If there is no dispute, then an Adjudicator has no jurisdiction to consider the matter.

As an example, if the contractor notifies a compensation event and the Service Manager does not agree with it, then the simple fact that the matter is not agreed does not mean that there is necessarily a dispute as the parties are still free to meet and, if necessary, to negotiate a settlement. It is only when the parties have exhausted all the opportunities to meet and reach a compromise, but still have a difference, that a dispute exists, and the matter can be referred to the Adjudicator.

It must also be noted that an Adjudicator only has jurisdiction to consider one dispute at a time under the same contract unless there is agreement between the disputing parties. Therefore, if there are number of issues in dispute, unless they are inextricably joined then there is likely to be several adjudications.

Disputes under or in connection with the contract may be referred to the Adjudicator.

Note that a dispute under the contract cannot be referred to the tribunal (arbitration or litigation) <u>unless</u> it has first been decided by the Adjudicator.

Adjudication is a quicker, convenient and less expensive way of resolving a dispute, whereby an impartial third party Adjudicator either named in the contract, selected by agreement between the parties or nominated by a third party, decides the issues in dispute between the parties. It is quicker and less expensive than arbitration or litigation, though the parties may refer to one of those methods following adjudication.

It must be recognised that the role of the Adjudicator is to <u>enforce</u> the contract, not to decide what, in their opinion, would be the fairest outcome. For example, if the parties have a dispute over the quality of the replacement of suspended ceiling tiles carried out, it is the Adjudicator's task to ascertain whether the tiles were supplied and installed in accordance with the contract, i.e. specifications, etc. and the applicable law, not whether it was a good or bad job.

10.12   Adjudication provisions within the TSC

As stated above, the TSC provides two Options for referral to the Adjudicator:

(i) <u>Option W1</u>

Option W1 is used when the UK Housing Grants, Construction and Regeneration Act 1996 does not apply.

Any dispute arising under the contract is firstly referred by a Party issuing a notice to the Senior Representatives named in Contract Data Parts 1 and 2, and copied to the other Party and the Service Manager stating the nature of the dispute it wishes them to resolve.

Each Party then submits their statement of case (no more than ten sides of A4 paper) with supporting evidence.

The Senior Representatives then try to resolve the dispute within no more than three weeks, using any procedure they wish to. At the end of the three weeks, a list is issued showing the issues agreed, and the issues not agreed. The Service Manager and the Contractor put the agreed issues into effect.

The Dispute Resolution Table includes the following:

If the Dispute is about. . . . . .

- An action or inaction of the Service Manager, either Party may refer it to the Senior Representatives, not more than four weeks after the Party became aware of the action or inaction.
- A plan, programme, compensation event, or quotation for a compensation event which is treated as having been accepted, the Client may refer it to the Senior Representatives, not more than four weeks after it was treated as accepted.
- An assessment of Defined Cost which is treated as correct, either Party may refer it to the Senior Representatives, not more than four weeks after the assessment was treated as correct.
- Any other matter, either Party may refer it to the Senior Representatives, when the dispute arises.

Note that if the referring Party is the Client and the dispute is regarding a compensation event which has been treated as having been accepted (Clause 62.6), then the Client must notify the dispute to the Contractor and then refer it to the Adjudicator not more than four weeks after it was treated as accepted.

This seems an odd reason for referring a dispute to the Adjudicator, as Clause 62.6 states that if the Service Manager does not reply to the Contractor's notification it is treated as acceptance of the quotation, how could an Adjudicator whose role is to enforce the contract find other than in favour of the Contractor? Surely the Client's dissatisfaction, and the dispute itself is with the Service Manager, not with the Contractor? But that is what the contract says, so the Adjudicator has the authority to open up a compensation event that has been accepted in that manner.

For any other matter which the Client wishes to refer to the Adjudicator, it may be referred between two and four weeks after notifying the Contractor and the Service Manager.

While these time scales are fixed within the Adjudication Table, they can be extended by the Service Manager if the Contractor and the Service Manager agree before either the notice or the referral is due. Note that if the matter in dispute is not notified and referred within these time scales, neither Party can refer the dispute to the Adjudicator or the tribunal.

The Adjudicator is appointed under the NEC4 Dispute Resolution Service Contract.

The Adjudicator acts impartially, and if they resign or are unable to act, the Parties jointly appoint a new Adjudicator. The referring Party obtains a copy and completes the Dispute Resolution Service Contract. If the Parties have not appointed an Adjudicator, either party may ask the nominating body to choose an Adjudicator.

No procedures have been specified for appointing a suitable person, and in practice a number of different methods have been used. Whatever method is used, it is important that both Parties have full confidence in the Adjudicator's impartiality, and for that reason it is preferable that a joint appointment is made.

The Adjudicator should be a person with experience in the type of work included in the contract between the Parties and who occupies or has occupied a senior position dealing with disputes. It should be able to listen and to understand the viewpoint of both Parties.

Often the Parties delay selecting an Adjudicator until a dispute has arisen, although this frequently results in a disagreement over who should be the Adjudicator.

As noted, the selection of the Adjudicator is important, and it should be recognised that a failure to agree an Adjudicator means that a third party will make the selection without necessarily consulting the Parties.

The referring Party must include within its referral information which it wishes to be considered by the Adjudicator, any more information from either Party to be provided within four weeks of the referral.

A dispute under a subcontract, which is also a dispute under the TSC may be referred by the Contractor to the Adjudicator at the same time, and the Adjudicator can decide the two disputes together if the Subcontractor agrees.

The Adjudicator may review and revise any action or inaction of the Service Manager, take the initiative in ascertaining the facts, and the law relating to the dispute, instruct a party to provide further information and instruct a party to take any other action which it considers necessary to reach its decision.

All communications between a Party and the Adjudicator must be communicated to the other Party at the same time.

The Adjudicator decides the dispute and notifies the Parties and the Service Manager within four weeks of the end of the period for receiving information.

This four weeks period may be extended by joint agreement between the Parties.

Until this decision has been communicated, the Parties proceed as if the matter in dispute was not disputed.

The Adjudicator's decision is binding on both of the parties unless and until revised by a tribunal and is enforceable as a contractual obligation on the parties. The Adjudicator's decision is final and binding if neither Party has notified the Adjudicator that they are dissatisfied with an Adjudicator's decision within the time stated in the contract, and intends to refer the matter to the tribunal.

The Adjudicator may, within two weeks of giving its decision to the Parties, correct a clerical mistake or ambiguity.

Review by the tribunal

A dispute cannot be referred to the tribunal unless it has first been referred to the Adjudicator.

The tribunal may be named by the Client within Contract Data Part 1.

While no alternatives are stated, it will normally be litigation or arbitration. If arbitration is chosen the Client must also state in Contract Data Part 1, the procedure, the place where the arbitration is to be held, and the person or organisation who will choose an Arbitrator if the Parties cannot agree a choice, or if the named procedure does not state who selects the Arbitrator.

A Party can, following the adjudication, notify the other party within four weeks of the Adjudicator's decision that it is dissatisfied. This is a time barred right, as the dissatisfied Party cannot refer the dispute to the tribunal unless it is notified within four weeks of the Adjudicator's decision, failure to do so will make the Adjudicator's decision final and binding.

The tribunal settles the dispute and has the power to reconsider any decision of the Adjudicator and review and revise any action or inaction of the Service Manager.

It is important to note that the tribunal is not a direct appeal against the Adjudicator's decision, the parties have the opportunity to present further information or evidence that was not originally presented to the Adjudicator, and also the Adjudicator cannot be called as a witness.

(ii) Option W2

Option W2 is used when the British Housing Grants, Construction and Regeneration Act 1996 applies and also with similar legislation in Northern Ireland with the additional inclusion of the option Y(NI)1.

The inclusion of adjudication within the then New Engineering Contract, now the NEC4, pre-dates British legislation giving parties to a contract the statutory right to refer a dispute to adjudication.

For Option W2, there is no Adjudication Table as the Parties have a right to refer a dispute to each other and to the Adjudicator at any time.

Any dispute arising under the contract may first be referred by a Party issuing a notice to the Senior Representatives and copied to the other Party and the Service Manager stating the nature of the dispute it wishes them to resolve. But this stage in the dispute resolution is optional so as to maintain the ability for a party to adjudicate at any time in accordance with the Act.

Each Party then submits its statement of case (no more than ten sides of A4 paper) with supporting evidence.

The Senior Representatives then try to resolve the dispute within three weeks, using any procedure they wish to. At the end of the three weeks, a list is issued showing the issues agreed, and the issues not agreed. The Service Manager and the Contractor put the agreed issues into effect.

The requirements of the Housing Grants Construction and Regeneration Act 1996 for adjudication to be possible at any time means that the involvement of the Senior Representatives may be bypassed by parties and the dispute referred straight to the adjudicator.

Again, as with Option W1, the Adjudicator is appointed under the NEC4 Dispute Resolution Service Contract. The Adjudicator acts impartially, and if he/she resigns or is unable to act, the Parties jointly appoint a new Adjudicator. The referring Party obtains a copy and completes the Dispute Resolution Service Contract. If the Parties have not appointed an Adjudicator, either party may ask the nominating body to choose an Adjudicator.

A Party may first give a notice of adjudication to the other party with a brief description of the dispute, details of where and when the dispute has arisen, and the nature of the redress which is sought.

The Adjudicator may be named in the contract in which case the party sends a copy of the notice to the Adjudicator. The Adjudicator must confirm within three days of receipt of the notice that it is able to decide the dispute, or if it is unable to decide the dispute.

Within seven days of the issue of the notice of adjudication, the party:

- refers the dispute to the Adjudicator;
- provides the Adjudicator with the information on which it relies together with supporting information;
- provides a copy of the information and supporting documents to the other party.

Again, a dispute under a subcontract, which is also a dispute under the TSC may be referred to the Adjudicator at the same time and the Adjudicator can decide the two disputes together.

The Adjudicator may review and revise any action or inaction of the Service Manager, take the initiative in ascertaining the facts, and the law relating to the dispute, instruct a party to provide further information and instruct a party to take any other action which it considers necessary to reach its decision.

All communications between a Party and the Adjudicator must be communicated to the other party at the same time.

The Adjudicator decides the dispute and notifies the parties and the Service Manager within 28 days of the dispute being referred. Note, to comply with the Act, time periods in W2 are measured in days, not weeks.

This 28 days period may be extended by 14 days with the consent of the referring Party or by any other period by joint agreement between the parties.

Until this decision has been communicated, the parties proceed as if the matter in dispute was not disputed.

The Adjudicator's decision is temporarily binding on the parties unless and until revised by a tribunal and is enforceable as a contractual obligation on the parties.

The Adjudicator's decision is final and binding if neither party has notified the Adjudicator that they are dissatisfied with an Adjudicator's decision within the time stated in the contract, and intends to refer the matter to the tribunal.

The Adjudicator may, within five days of giving the decision to the parties, correct any clerical mistake or ambiguity.

Review by the tribunal

A dispute cannot be referred to the tribunal unless it has first been referred to the Adjudicator.

Again, the tribunal may be named by the Client within Contract Data Part 1, normally litigation or arbitration. If arbitration is chosen the Client must also state in Contract Data Part 1, the procedure, the place where the arbitration is to be held, and the person or organisation who will choose an Arbitrator if the Parties cannot agree a choice, or if the named procedure does not state who selects the Arbitrator.

A party can, following the adjudication, notify the other party within four weeks of the Adjudicator's decision that it is dissatisfied. This is a time-barred right, as the dissatisfied Party cannot refer the dispute to the tribunal unless it is notified within four weeks of the Adjudicator's decision, failure to do so will make the Adjudicator's decision final and binding.

The tribunal settles the dispute and has the power to reconsider any decision of the Adjudicator and review and revise any action or inaction of the Service Manager.

It is important to note that the tribunal is not a direct appeal against the Adjudicator's decision, the parties have the opportunity to present further information or evidence that was not originally presented to the Adjudicator, and also the Adjudicator cannot be called as a witness.

10.13   NEC4 Dispute Resolution Service Contract

Introduction

The NEC4 Dispute Resolution Service Contract replaced the NEC3 Adjudicator's Contract.

The first edition of the NEC Adjudicator's Contract was published in 1994, and was written for the appointment of an Adjudicator for any contract under the NEC family of standard contracts. The second edition, published in 1998, contained some changes including the need to harmonise with the NEC standard contracts and further editions which had been issued since 1994. The third edition harmonised the contract with the NEC3 family of contracts using either Option W1 or W2.

The Dispute Resolution Service Contract is made up of five parts:

1. General
2. Adjudication
3. Dispute Avoidance Board
4. Payment
5. Termination

The Dispute Resolution Service Contract can also be used with contracts other than NEC4, though dependent on the contract and the applicable law, some amendment may be necessary.

In the UK, the Housing Grants, Construction and Regeneration Act 1996 has made adjudication mandatory as a means of resolving disputes in construction contracts which fall under the Act. Parties to a contract which does not provide for adjudication as required by the Act have a right to adjudication under the "Scheme for Construction Contracts". Schemes which are substantially similar have been published for England, Wales, Scotland and Northern Ireland.

The agreement is between the two disputing parties and the Adjudicator when used with the TSC.

The contract contains five sections:

1.   General

This covers the obligation upon the Dispute Resolver to act impartially, and to notify the parties as soon as it becomes aware of any matter which could present a conflict of interest.

There is a definition of expenses, which includes printing costs, postage, travel, accommodation and the cost of any assistance with the adjudication.

All communications must be in a form which can be read, copied and recorded. This is a requirement of all NEC contracts, but also of the adjudication process itself, with both Parties and the Dispute Resolver having to be copied into any communication between the parties.

2.   Adjudication

This Clause only applies if the Dispute Resolver is acting as an Adjudicator.

The Dispute Resolver cannot decide any dispute that is the same or substantially the same as its predecessor decided. It must make a decision and notify the Parties in accordance with the contract, and in reaching its decisions it can obtain assistance from others, but must advise the Parties before doing so, and also provide the Parties with a copy of anything produced by the assisting party, so that the Parties can be invited to comment.

The Dispute Resolver's decision is to remain confidential between the parties, and following its decision the Dispute Resolver retains documents provided to them for the period of retention, which is stated in the Contract Data.

An invoice is issued by the Dispute Resolver:

* each time a dispute is referred if an advance payment is stated in the Contract Data
* after a decision had been notified to the Parties
* after a termination

3.   Dispute Avoidance Board

This Clause only applies if the Dispute Resolver is acting as a Dispute Avoidance Board member under other NEC4 contracts, but not the TSC.

The Dispute Resolver carried out the duties of a Dispute Avoidance Board member in accordance with the contract between the Parties and collaborates with other members of the Dispute Avoidance Board.

4.   Payment

The Dispute Resolver's hourly fee is stated in the Contract Data. There is provision for an advance payment to be made by the referring party to the Dispute Resolver if stated in the Contract Data.

Following the decision, the Dispute Resolver submits an invoice to each Party for their share of the amount due, including expenses. The Parties pay the amount due within three weeks of receiving the Dispute Resolver's invoice. Interest is paid at the rate stated in the Contract Data on any late payment. The Parties are jointly and severally liable to pay the amount due to the Dispute Resolver, therefore if one party fails to pay, the other must pay including any interest on the overdue amount, and recover the amount from the other Party.

5.   Termination

The Parties may, by agreement, terminate the Dispute Resolver's agreement. Also, the Dispute Resolver may terminate the agreement if there is a conflict of interest, it is unable to decide a dispute, an advance payment has not been made, or a payment of an amount due has not been made within five weeks of the date by which the payment should have been made. The Dispute Resolver's appointment terminates on the date stated in the Contract Data.

10.14   The tribunal

While Contract Data Part 1 requires the Client to enter the form of the tribunal, it does not say what choices are available. Other contracts refer disputes to expert determination or Dispute

Adjudication Boards, but under the TSC the Client normally selects arbitration or litigation. Later, within the Contract Data the Client is required to state the arbitration procedure, the place where the arbitration is to be held, and who will choose an Arbitrator.

There are a number of differences between the two, but essentially arbitration is hearing the dispute in private, and litigation is hearing the dispute in a public court.

It must be stressed that the role and authority of the tribunal is not to review and appeal against the previous decision of an Adjudicator, but to resolve the dispute independent of the adjudication, therefore it is quite in order for parties to introduce new evidence or materials to the tribunal. The courts in the UK have largely supported adjudicator's decisions and typically only refuse to when errors of process, fairness or bias have occurred.

10.15   Arbitration

Any dispute between two or more parties can be resolved through the courts, a process known as litigation.

However, litigation has many disadvantages, not least the cost of a full court hearing, but in some cases a lengthy waiting time before the matter actually reaches the courts, so arbitration has become increasingly used as a simpler, more convenient method of dispute resolution.

Arbitration is the hearing of a dispute by a third party, who is often not a lawyer, though that term has a very wide meaning, but is an expert in the field in which the dispute is based, and can give a decision based on opinion, which is then legally binding on both parties.

Arbitration is not a new concept, in fact it has been in existence for almost as long as the law itself, the first official recognition in the UK being the Arbitration Act 1697, which largely governed disputes about the sale of livestock in cattle markets. With the continuing growth of freight transport by ship, and the advent of railways and other forms of transport, a large number of arbitration cases ensued, and as a result Parliament passed the Arbitration Act 1889.

Various amendments have been introduced over the following years with the current legislation being the Arbitration Act 1950, as amended by the Arbitration Acts of 1975 and 1979. The Arbitration Act 1996 is the current legislation related to arbitration in England and Wales.

One does not issue writs in arbitration as one would with litigation, *both* parties agree to enter into arbitration as a pre-agreed provision within a contract and thus be bound by the decision of the Arbitrator.

Arbitrators are appointed by one of THREE methods:

(i)   As a result of a pre-agreed decision between the parties, normally though a provision within a contract to go to arbitration in the event of a dispute.

   In this case the Arbitrator himself may be named or subject to nomination by a professional body such as the Chartered Institute of Arbitrators.

(ii)   By Statute, legislation in many countries includes the provision for disputes to be settled by arbitration.

(iii)   By Order of a Court.

The Arbitration Act is fairly brief and sets out procedures in the absence of any agreement to the contrary between the parties.

As arbitration is a more flexible arrangement than litigation, whatever procedure the Arbitrator and both parties agree is sufficient in an arbitration case.

While most forms of contract provide for resolution of disputes by arbitration, it must be stressed that it should only be seen as a last resort when all attempts at negotiation and other resolution methods have failed.

It may be surprising to hear that arbitration is common as a settlement procedure for disputes under trade agreements, maritime and insurance, consumer matters such as package holidays, construction industry disputes and property valuations.

The contracting party that initiated that the dispute be referred to arbitration is always referred to as the 'Claimant', while the other party is referred to as the 'Respondent'. Should there be a Joint Agreement to go to arbitration, the Arbitrator will decide who is the Claimant, and who is the Respondent.

It is important that these titles are clarified as the Arbitration Rules repeatedly refer to them. There are several <u>advantages</u> in using arbitration rather than litigation:

(i) Privacy

This is a distinct advantage, as in litigation, private and often personal matters are debated in full view of the public in open court, and often published in the press and technical magazines, whereas arbitration is carried out in private sessions, usually at the Arbitrator's office or at a pre-arranged venue.

(ii) Convenience

Arbitration is carried out at a time and place to suit the parties, for example evenings and weekends, not when the court is available for session.

(iii) Speed

The dispute is settled efficiently without the normal delays and procedural matters involved in often fairly lengthy and formal litigation.

(iv) Simplicity

There are few technical procedures as in the courts, the parties set their own procedures within a set of ground rules.

(v) Expertness

The Arbitrator will be an expert on the matter in dispute, for example a structural engineer where the dispute is about structural design issues, whereas judges and other legal professional are experts only on principles of law. The dispute will often not hinge on a legal problem, for example in matters of design who better than a designer to resolve it?

(vi) Thoroughness

Because the Arbitrator is a skilled technical person, the dispute can often be dealt with more thoroughly and matters more rigorously aired than in a confined and formal court atmosphere.

(vii) Expense

It can be readily seen that arbitration is generally a less expensive option than litigation, though it must not be seen as a cheap way to resolve a dispute. In some complex cases arbitration can be even more expensive than litigation.

It is worth mentioning at this point that, in many countries, legal aid, which can be available for litigation, is not available for arbitration.

The main <u>disadvantages</u> of using arbitration rather than litigation are:

(i) Lack of legal knowledge

A dispute may hinge on a difficult point of law, which the Arbitrator has insufficient legal knowledge to resolve, as it is an expert in the field relating to the dispute, not in the law itself. It is, however, entitled to seek legal advice on a point of law, but not on the dispute itself, and also the disputing parties can refer a question of law to the High Court.

NB: The arbitration award itself cannot be referred to the Courts unless both parties consent, or the Courts are referred to on a point of law, the resolution of which could substantially overturn an Arbitrator's decision.

Furthermore, Arbitrators cannot be sued for negligence as they do not hold themselves up as experts in the sense of being advisory bodies and therefore liable for giving wrongful advice.

(ii) Precedents

Arbitration is not subject to rulings in previous cases as with the courts, each case being judged on its own merits on the day, and also to some extent subject to the personality of the Arbitrator, therefore there is no guide as to how successful an action could be.

It is prudent to always seek professional advice before resorting to arbitration as while it is simpler than Court action the Arbitrator's decision is final and binding, therefore a party must be confident that they have a strong case.

(iii) Decision-making

It has been said that as the Arbitrator, being a technical person and possibly having been involved in a similar dispute themselves, tends to be fairer than judges in litigation, they also tend to sympathise with both parties and can split the difference between the parties when making their awards as they have possibly been in a similar position themselves, but judges are inclined to take colder decisions and award wholly in favour of one of the parties. It has to be said that this is not a commonly held view!

Reference to arbitration

Although arbitration is a different process to adjudication, the same initial principles apply before arbitration can commence in that there must be a dispute, there must be an agreement to arbitrate, which will normally be provided for within the contract, and thirdly a party must submit a notice to initiate the process.

Proceedings begin by either party sending a written notice to the other stating that they wish to refer the dispute to arbitration, the form and process may be dictated by the arbitration procedure stated by the Client within Contract Data Part 1.

The parties may agree on a person to act as Arbitrator, or a professional body will select and appoint the Arbitrator. In some cases, where the parties cannot agree a choice and there is no provision for a third party to do it, a court may select an Arbitrator. There is normally only one Arbitrator, but in some cases three, consisting of two Arbitrators and a third who acts as an umpire. Any correspondence flowing between the two parties as from the date of appointment must then also be copied to the Arbitrator(s).

The next step is that the Arbitrator writes to the parties accepting the appointment, a Preliminary Meeting is held between the parties at a time and venue to be directed by the Arbitrator to decide the form of the hearing which may be on a documents only basis or an oral hearing. Documents only tends to be cleaner and quicker as there is less opportunity for the parties to debate the issue.

The role of the Arbitrator

The role of the Arbitrator is only to examine the evidence before them enabling them to resolve the dispute, and nothing more, it cannot look into other matters not in dispute. It can conduct the arbitration as it wishes, though it must gear the procedures in a way that is not detrimental to either party.

It must also have a regard for the rules of evidence.

For example:

(i) Hearsay evidence is not permissible.
(ii) It is not permitted to use evidence from someone not qualified to give such evidence.
(iii) The Arbitrator is not allowed to take into account any evidence it has discovered itself, which was not provided by either party.
(iv) It is not allowed to use any privileged evidence, for example that which is marked 'Without Prejudice'.
(v) In addition, the rules of discovery apply where either party is entitled to have access to the file or documents of the other provided that it relates to matters at issue.

Inspection by the Arbitrator

The Arbitrator is entitled to inspect any property, work or material at any premises, and may invite the Claimant and/or the Respondent to accompany them purely for the purpose of identifying the work or materials.

No-one else is allowed to accompany the Arbitrator unless specifically invited by the Arbitrator.

The award

The award is binding upon both parties and must be served in the appropriate manner with full headings, summary of issues in dispute, outline of events leading to the dispute, its decision based on the hearing and finally the monetary award.

There is very limited scope for appeal against an Arbitrator's award. Usually, appeals can only be based on a claim that the Arbitrator was wrong on a point of law, or there was serious and proven irregularity.

Arbitration awards can also be enforced in court if necessary.

Arbitrator's fees and expenses

Both parties will normally incur costs, though it is the Arbitrator's decision as to how the costs are apportioned. Both parties are initially liable to the Arbitrator for payment of fees and expenses, but the Arbitrator will direct the parties as to the proportion they are liable for once the decision is reached.

Should the parties agree on a settlement of the dispute at any time during the arbitration, they will jointly be liable for the Arbitrator's costs.

10.16   Litigation

It is not proposed to go into any detail about the litigation process from initiation to judgment by a court, first as the contract does not give any details of the process, but also the proceedings themselves will depend on the law of the contract, which in turn is dependent on the location of the service.

However, the process will normally involve the Claimant's representative, usually a Solicitor issuing a claim form or write to the appropriate court, papers are then served on the Defendant. The Defendant then replies to the service by either admitting or denying the claim against it, if denied, it replies by sending its defence which may also be accompanied by any counter-claim. The case is then allocated to a court. There then follows a period of accumulation, collation and presentation of evidence, and hearings, again the process, the submission and the form of the hearing being dependent upon the law of the contract, until the court passes judgement. Cases in most British courts will normally see barristers acting as advocates on behalf of the parties alongside the solicitors who have advised the parties.

11 Tenders

11.1 Deciding the procurement strategy

Procurement strategy considers the most optimum way of achieving the client's objectives, taking into account its current and future needs, and the inherent risks in achieving those objectives.

Deciding the procurement strategy is the foundation stone to a successful outcome to a contract, and a well-considered and appropriate procurement strategy together with a clear, robust and unambiguous tender document is the key to avoiding or minimising contractual problems at a later date.

But this process is very often overlooked as clients and their advisers, in their urgency to get projects and services out to tender and commenced, tend to favour the procurement strategy most familiar to them, rather than to spend time properly considering the project or service in terms of the client's resources and its objectives in terms of time, cost and quality, with the additional factor of risk having to also be considered and evaluated. It is rarely achievable for all three to be fully satisfied it is usually a balance.

Clients should always carefully select a procurement strategy most appropriate to their needs, the project or the service, and the various risks involved. The procurement strategy should consider value for money in a holistic way, considering life cycle costing as well as the capital cost of building the project or providing the service, though it is not intended to consider life cycle costs in any detail within this book.

The procurement strategy then leads to a contract strategy which should be capable of achieving the aims of the procurement strategy.

While many traditional contracts for services are based on price lists with the client specifying the service, and outline price lists being sent to each tendering contractor for pricing, in the past 20 years there has been a movement away from the use of traditional bills of quantities and in terms of service contracts, which we are discussing within this book, towards payment mechanisms such as milestone payments and activity schedules, with payment based on progress achieved, rather than quantity of work done. With service-based contracts, it is rare to use bills of quantities anyway.

The initial briefing is the fundamental first step in establishing what is required, when it is required, and what the client is proposing, or sometimes can afford to pay.

Through a process of interviewing key members of the client's team and reviewing its business plans, specific objectives, options and constraints, a procurement strategy is developed which forms the basis for future assessments and recommendations.

While it may sound obvious, it is vital that the client understands what they want and that they can convey it clearly to others. The client may not know definitively what it wants, but also a client can often be swayed by persuasive consultants or other third parties.

DOI: 10.1201/9781003463771-12

Time

All clients want their services to be carried out efficiently, and in good time. This is particularly relevant for example in the retail sector where the client needs its stores properly maintained and open for trading each morning.

If the client sees time as being the most important factor, it may have to pay more money and quality may lower in order to achieve that.

Price

All clients will have some form of budget, so this will always be a major consideration. The budget may be fixed, representing all that the client has available in order to fund the service, or may have some degree of flexibility. The form of procurement will need to address the budget and any inherent flexibility.

Quality

Finally quality will always be a major factor, though if the price is important, some clients will lower the expected standards.

Once these have been established and analyzed, then the appropriate decisions can be made, and within the TSC on Main and Secondary Options and the allocation of responsibilities and the timing for preparing the relevant documents.

Contractor design

The TSC does not have express provisions for contractor design of any permanent works, though some clients include clauses to provide for it.

In reality, there tends to be very little contractor design when using the TSC, though Task Orders can often include for physical construction work partially or totally to be designed by the Contractor.

The authors are aware of a particular TSC where the Contractor providing services to government buildings in London, was required to design and install a lift within an existing building as a Task Order!

There are a number of reasons for allocating some or most of the design to the contractor:

• The design and construction periods can overlap, leading to faster delivery of the project.
• The contractor can utilise its experience and preferred methods of construction to build in a rational way which minimises costs and programme durations.
• The temporary works/permanent works interface and influence of design is rationalised as both remain with the contractor and therefore the permanent works should be more "buildable".
• The traditional design/construction interface and the risks associated with it are transferred to the contractor.
• The management of the design risk by the contractor can result in greater certainty of the time, cost and performance and project objectives being met.
• The Contractor is well-placed to integrate its design, construction and maintenance services.

There has also been an increasing use in the past 20 years of target contracts which have been encouraged by the increasing use of partnering and collaborative working arrangements, and the associated more equitable sharing of risk. It has to be said, that the NEC contracts have also played a great part in increasing the use of target contracts

Through its Main and Secondary Options, the TSC caters for all the procurement methods commonly used in service contracts throughout the world.

Factors to take into account in deciding which procurement option to use include the following:

- How important to the client is certainty of price, particularly if changes are expected? Lump sum options will give greater certainty of price.
- What resources and expertise does the client have?
- As stated above, which party is to be responsible for design and/or which party has the necessary design expertise? Design can be assigned to the contractor under any of the Main Options, though as stated above, it is rare for design to be assigned to the contractor in term service contracts.
- How important to the client is an early start and/or early completion of tasks? Early commencement and early completion may lead the client toward a cost reimbursable solution as lump sum and remeasurement procurement options require a considerable time to prepare the tender documentation for pricing.
- How clearly defined is the Scope, and what form does it take, for example is it in the form of a specification, or performance-based criteria? Again, the contracts can be used with detailed specification or performance-based criteria.
- What is the likelihood of change to those defined requirements? Many clients prefer a remeasurement option where there is likelihood of many changes.
- What is the client's attitude toward risk? Who is best placed to manage the inherent risks within the service? Lump sum contracts assign greatest risk to the contractor, cost reimbursable contracts assign greater risk to the client. It is a poorly acknowledged fact that the client always pays for risk, regardless of whether it is carried within the contract by the client or contractor. If it is the contractor's risk it is deemed to have priced and programmed for it, if it is the client's risk then it will be paid for if it occurs.

11.2 Partnering arrangements

The past 30 years, particularly since the publication of the Latham and Egan reports have seen the growth of partnering and framework agreements.

The US Construction Industry Institute defines partnering as:

"A long-term commitment between two or more organisations for the purpose of achieving specific business objectives by maximising the effectiveness of each participant's resources . . . the relationship is based upon trust, dedication to common goals and an understanding of each other's individual expectations and values".

Partnering is a medium to long term relationship between contracting parties, whereby the contractor and in turn consultants and various other parties are not required to tender for each project, but is awarded the work, normally by negotiation.

The advantages of partnering agreements are:

- Reduced costs of rebidding for future contracts
- Developed relationships based on trust
- Contractors are appointed earlier and can contribute to the procurement process
- Greater cost certainty

- Continuous improvement by transferring learning from one contract to another
- Improved working relationships
- Continuous workflow
- Speed of procurement

The risks within a contract are initially owned by the client, who may choose to adopt a "risk transfer" approach where the risks are assigned through the contract to the contractor who has the opportunity to price and programme for them, or a "risk embrace" approach where the client retains the risks. In reality, most contracts are a combination of the two. The traditional approach to risk management is that of risk transfer, which is fine if the scope of work is clear and well defined, however, in recent years clients have become more aware that they can achieve their objectives better by adopting a more "old fashioned" risk embrace culture.

The TSC includes a Secondary Option (X12) for multiparty collaboration. This is the Option which, if included in NEC contracts, creates obligations between the Partners listed in the Contract Data. The content of Option X12 is originally derived from the "Guide to Project Team Partnering" published by the Construction Industry Council

11.3 Provisional and Prime Cost (PC) Sums

Provisional and PC Sums are used in other contracts where there are elements of work which cannot be sufficiently defined at the time of tender and therefore a sum of money is included by the client within the pricing documents to cover the item. When the item is defined or able to be properly defined the contractor is given the information which allows for pricing, the Provisional Sum is omitted and the rice included in its stead.

The problem with Provisional Sums is that they reduce the competition among tenderers as they are not priced at tender stage, and also if they are "defined" the contractor is deemed to have allowed time in the programme for them, if they are "undefined" it has not.

The NEC contracts have never provided for Provisional and PC Sums, the principle being that the client decides what they want at tender stage so it can be designed and accurately described and the tendering contractors can properly price and programme for them, or when the client does decide what it wants the Service Manager can give an instruction to the contractor which changes the Scope, it is a compensation event under Clause 60.1(1) and can be priced at the time. In that case, the cost of the work is held in the budget for the scheme rather than in the contract.

11.4 The pre-tender planning stage

Assuming a traditional main contract, the contractor selection process follows the following steps:

- Compiling the shortlist
- Criteria for contractor selection
- Information for contract tender enquiries
- Comparison of tenders
- Contract Award

Each of the above stages will be dealt with in the form of a checklist.

11.5   Preparing tender documents

The tender documents for a TSC contract will normally consist of:

(i)   Invitation to Tender (including instructions to tenderers on time and place of tender submission)
(ii)   Form of Tender
(iii)   The Pricing Document i.e. a Price List for an Option A or C contract
(iv)   The Scope
(v)   Contract Data Part 1 (completed by the client)
(vi)   Contract Data Part 2 (blank pro forma to be completed by the tendering contractors)

In addition, there may be other information, dependent on the applicable legislation, for example in the UK, where the Construction (Design and Management) Regulations 2015 apply, Pre-Construction Information would be included within the tender documents.

11.6   European Union procurement directives

While not wishing this book to be geographically specific, and the UK has left the EU, it is worth for readers within the EU some consideration of the European Union (EU) procurement directives that currently apply to all procurement within the public and regulated sectors, above minimum monetary thresholds which are reviewed on a regular basis, and subject to EU-wide principles of non-discrimination, equal treatment and transparency. The regulations affect government departments, local authorities and health authorities, and also utilities companies operating in the energy, water and transport sectors.

The purpose of the procurement rules is to open up the public procurement market and to ensure the free movement of supplies, services and works within the EU.

The rules are enforced by the Members States' Courts and the European Court of Justice and require that all public procurement must be based on value for money, defined as "*the optimum combination of whole-life cost and quality to meet the user's requirement*", which should be achieved through competition, unless there are compelling reasons to the contrary.

In addition to the EU Member States, the benefits of the EU public procurement rules also apply to a number of other countries outside Europe because of an international agreement negotiated by the World Trade Organisation titled the "Government Procurement Agreement".

The EU Procurement Directives set out the legal framework for public procurement when public authorities and utilities seek to acquire supplies, services, or works (e.g. civil engineering or building), procedures which must be followed before awarding a contract when its value exceeds set thresholds which are reviewed on a regular basis, unless the contract qualifies for a specific exclusion, for example on grounds of national security.

The Directives require contracting authorities to provide details of their proposed procurement in a prescribed format, which are then published in the *Official Journal of the European Union* (OJEU).

All companies replying to an OJEU advertisement have an equal opportunity to express interest in being considered for tendering. As in all tendering exercises, clients must ensure that those companies selected to tender receive exactly the same information on which to make their bid.

The notice may be issued electronically and this service may be provided through an intermediary organisation.

Generally contracts covered by the Regulations must be the subject of a call for competition by publishing a Contract Notice in the OJEU.

The OJEU Notice process has associated minimum timeframes for receipt of tenders or requests to participate, dependent on procurement procedure, from the date the contract notice is sent.

11.7   Stages in the procurement process

The Regulations set out criteria designed to ensure all suppliers or contractors established in countries covered by the rules are treated on equal terms, to avoid discrimination on the grounds of origin in a particular Member State.

The criteria cover:

- Specification stage: how requirements must be specified
- Selection stage: the rejection or selection of candidates
- Award stage: the award of contract, either on the basis of "lowest price" or preferably the "most economically advantageous tender" to the purchaser

11.8   Invitation to tender

As soon as it has been decided to select a contractor by competitive tender, a shortlist of those considered suitable to be invited to tender should be compiled.

It is often the case that clients and their representatives already hold a list of contractors of established skill, integrity, responsibility and proven competence. This should be considered as a matter of policy within the client's organisation.

Clients should review their lists periodically so as to exclude firms whose performance has been unsatisfactory in the past, and to allow the introduction of new firms.

The cost of preparing tenders is a significant element of the overheads both of the contractors and in turn their subcontractors. Tender lists therefore should be kept as short as practicable. Enquiries should be kept to between three and six, depending upon the type of work and size of project.

The use of a register of contractors at the selection stage is recommended. This may lead to establishing contractor record cards relating to performance criteria on previous contracts.

It is important that such records are regularly updated and consideration given to adding new contractors to the register. It may be advisable at this stage to consider using a database for compiling the register. This may prove advantageous to a medium to large sized organisation in order to speedily access contractor information.

It is important that close relationships are encouraged between the client and the contractor, and in turn its subcontractors, for the prospering of both parties in the long term.

11.9   Information for tender enquiries

It is critical that clients properly identify what information should be contained in the enquiry to contractors.

Checklist for invitation to tender

- Details of services to be provided
- Service/project title and location of Affected Property

- Name of client
- Names of Service Manager and other consultants
- Details of the form of contract
- Scope – including general description of the service and specification
- Details of where further documents may be inspected
- Time period for completion of work (if known)
- Name of the Adjudicator (in case of dispute)
- Contractor's responsibility for accommodation and welfare arrangements and facilities:

  - Security
  - Storage facilities
  - Taking delivery of materials and consumables
  - Scaffolding and other temporary structures where applicable
  - Water and temporary electrical supplies
  - Safety, health and welfare provisions
  - Licences and permits
  - Any additional facilities
  - Access and security

Criteria for Contractor Selection

- Client's previous experience with the contractor
- The contractor's ability to manage its resources and liaise with the client's staff. Good relationships between parties are an essential requirement to developing a team approach to a successful project
- Financial standing of the contractor
- The contractor's expertise which they can bring to the contract
- The contractor's reputation and its standing with the client
- The current commitment of the contractor's organisation
- The contractor's current workload with other clients should be determined and serious consideration given to their ability to cope with the increased work.

A large number of contractors just cannot say 'No' when it comes to taking on more work. They often pull and push their limited workforce between sites hoping the client will not notice that they are stretched to the limit.

- The acceptability of the contractor to the client. On many contracts the contractor is required to name its subcontractors at the tender stage.
- The competitiveness of the contractor's price. The price must be accurate and reasonable otherwise the contractor will never win any work. Price discounts which may be applicable and the contractor's response to negotiation may be an important factor.
- The contractual risk which the contractor is expected, and prepared to take.
- The ability of the contract organisation to meet quality assurance criteria as laid down by the contract or as specified by the client.
- References available from the contractor. These include trade and bank references. The willingness of the contractor to allow previous contract work to be inspected. It is important that good relationships are established between the client and the contractor as early in the planning process as possible. This is especially important where the main contractor intends to sublet all the work on a particular project.

Any invitation to tender will give a brief description of the work, who it is for, and how tenders may be submitted.

Other inclusions will cover such matters as:

* Time, date and place for the delivery of tenders
* What documents must be included with the tender, including a programme
* The policy regarding on alternative and/or non-compliant bids
* Arrangements for visiting the site including contact details
* Rules on non-compliant bids
* Anti-collusion certificate

11.10   Form of Tender

There is no "pro forma" Form of Tender within the NEC contracts. This is the tenderer's written offer to execute the work in accordance with the contract documents.

The Form of Tender is normally in the form of a letter with blank spaces for tenderers to insert their name and other particulars, total tender price, and other particulars of their offer. It is essential to have a standard Form of Tender so that all tenderers utilise the same form, to make comparison of tenders easier.

11.11   The pricing document

The pricing document will depend on the Main Option chosen.

With Options A and C, the client is required to send the tenderers a blank Price List for completion. This comprises a list of items giving the quantities and a description of the work included in the contract and, in conjunction with the other contract documents, forms the basis upon which tenders are obtained and on which the contractor is paid.

11.12   Scope

Scope is defined by Clause 11.2(14) as "*information which:*

* *either specifies and describes the service; or*
* *states any constraints on how the contractor Provides the Service*

and is either:

* *in the documents which the Contract Data states it is in; or*
* *in an instruction given in accordance with this contract*".

The client provides its Scope and refers to it in Contract Data Part 1, if the contractor provides Scope, this is included in Contract Data Part 2.

The main documents within the Scope are normally the specifications, but will also include:

Description of the services

* a statement describing the scope of the service
* schematic layouts, plan, elevation and section drawings, detailed working and/or production drawings (if relevant and available), etc.
* a statement of any constraints on how the contractor Provides the Service, e.g. restrictions on access, sequencing or phasing of works, security issues

Plant and Materials

- materials and workmanship specifications
- requirements for delivery and storage
- provision of spares, maintenance requirements, etc.

Documentation required from the contractor as part of the service

- "As built" drawings
- Maintenance manuals
- Training documentation
- Test certificates
- Statutory requirements and/or certification

Health and safety

- Specific health and safety requirements for the Service Areas which the contractor must comply with, particularly if the site is within existing premises, including house safety rules, evacuation procedures, etc.
- Any information and health and safety plans for the service

Financial records

- Details of the accounts and records to be kept by the contractor (particularly under Options C and E)

Services

- Details of other contractors and any other parties who will be occupying the Service Areas during the service period and also any sharing requirements

Subcontracting

- Lists of acceptable subcontractors for specific tasks
- Statement of any work which should not be subcontracted
- Statement of any work which is required to be subcontracted

Plan

- Any information which the contractor is required to include in the plan in addition to the information shown in Clause 31.2. Also, if the client requires the contractor to produce a certain type of plan, paper/electronic copies, or to submit it using a particular software, then this should be clearly detailed within the Scope

Tests

- Description of tests to be carried out by the contractor and others
- Specification of materials, facilities and samples to be provided by the contractor and the client for tests
- Specification of Plant and Materials which are to be inspected or tested before delivery to the Service Areas
- Definition of tests of Plant and Materials outside the Service Areas which have to be passed

Others

- There are also certain specific requirements for statements to be made in the Scope from certain Main and Secondary Options in the conditions of contract.

Note that the TSC does not have Site Information, therefore any information regarding the Service Areas and/or Affected Property must be included within the Scope, for example:

(i) Information about existing buildings, structures and plant on or adjacent to the Service Areas including details of infrastructure such as lifts, escalators, boilers, ventilation plant and so on
(ii) Details of any previously demolished structures and the likelihood of any residual surface and subsurface materials
(iii) Reports obtained by the client concerning the physical conditions within the Service Areas or its surroundings. This may include mapping, hydrographical and hydrological information
(iv) All available information on the topography of the Service Areas should be made accessible to tenderers, preferably by being shown on the drawings
(v) Environmental issues for example asbestos, nesting birds, and protected species
(vi) References to publicly available information
(vii) Information from utilities companies and historic records regarding plant, pipes, cables and other services below the surface of the site

It is vital that care is taken to get the Scope correct.

Tenderers must be given sufficient information to enable them to understand clearly what is required and thus to submit considered and well-priced tenders.

Note that under Clause 63.11, if there is an ambiguity or inconsistency within the Scope, and it has to be changed to resolve the ambiguity or inconsistency, the contractor is assumed to have taken into account the conditions most favourable to doing the work in terms of Price and/or Task Completion Date. While many interpret that as the contractor allowing the cheapest way of doing the work, it may be the easiest or quickest way.

The Scope must be carefully drafted in order to define clearly what is expected of the contractor in the performance of the contract and therefore included in the quoted tender amount and plan. If the contract does not cover all aspects of the work, either specifically or by implication, that aspect may be deemed to be excluded from the contract.

If items of Plant and Materials are to be fabricated or manufactured off site by others, it is advisable that the contract sets out the corresponding obligations and liabilities of the respective parties, particularly if these are to form an integral or key part of the completed works.

The Scope describes clear boundaries for the work to be undertaken by the contractor. It may also outline the client's objectives and explain why the work is being undertaken and how it is intended to be used. It says what is to be done (and maybe what is not included) in general terms, but not how to do it or the standards to be achieved. It explains the limits, where the work is to interface with other existing or proposed facilities. It may draw attention to any work or materials to be provided by the client or others. It should also emphasise any unusual features of the services or contract, which tenderers might otherwise overlook.

This is the document that a tenderer can look to, to gain a broad understanding of the scale and complexity of the contract and be able to judge its capacity to undertake it. It is written specifically for each contract. In some respects it is analogous to a shopping list. It should be comprehensive, but it should be made clear that it is not intended to include all the detail, which is contained in the drawings, specifications and schedules.

The Scope may also include drawings which again should provide clear details of what the contractor has to do. Clearly, tenderers must be given sufficient information to enable them to understand what is required and thus submit considered and accurate tenders.

It will also include the specification which is a written technical description of the standards and various criteria required for the work, and should complement the drawings. The Specification describes the character and quality of materials and workmanship, for work to be executed.

Again, it may lay down the order in which various portions of the services are to be executed. As far as possible, it should describe the outcomes required, rather than how to achieve them. It is customary to divide the services into discrete sections or trades, with clauses written to cover the materials to be used, the packaging, handling and storage of materials (only if necessary), the method of work to be used (only if necessary), installation and maintenance criteria, the standards or tests to be satisfied, any specific requirements for completion, etc.

The specification is an integral part of the contract. This is often overlooked, with the result that inappropriate or outmoded specifications are selected, or replaced by a few brief notes on other contract documents. The contractor should spend an appropriate amount of time specifying the quality of the work, as it is not possible to price, build, test or measure the work correctly unless this is done.

The Scope describes what the contractor has to do in terms of scope and standards, and in some cases must not do or include in order to Provide the Service. It may also include the order or sequence in which the services are to be carried out. It also details where the work is to interface with other existing or proposed contractors or facilities. It will also include any work or facilities or materials to be provided by the client or others. It may also include any unusual features of the contract, which tenderers might otherwise overlook, for example any planning or operational constraints, etc.

This is the document that from which a tenderer can gain a broad understanding of the scale and complexity of the service and its capacity to undertake it.

It is written specifically for each contract.

The Scope should be:

- Clear: Unambiguous
- Concise: Not over wordy
- Complete: Have nothing missing

11.13   The Contract Data

The Contract Data provides the information required by the conditions of contract specific to a particular contract.

Part 1

Part 1 consists of data provided by the client, the sections of the Contract Data aligning with the sections of the core clauses and options.

Section 1: General

This section requires the client to identify the selected Main and Secondary Options, the service, and the names of the Client and the Service Manager.

While the person undertaking the role of Service Manager must always be a named individual, many clients choose to insert company names for these parties within the Contract Data and to separately identify the named individuals. It is also not uncommon for clients to name directors or partners of their respective companies, then the authority of the Service Manager is delegated to the individuals under Clause 14.2.

The Affected Property is also identified.

The Scope is also identified within this section, though this is normally incorporated by reference to separate documents rather than listing specification references, etc. If this is the case, the separate documents must be clearly defined.

The boundaries of the Affected Property are defined, sometimes in a list and sometimes by reference to a specific drawing or map.

As the NEC4 contracts are intended for use worldwide, the language and the law of the contract are also entered.

The period for reply is the period that the parties have to reply to submissions, proposals, notifications, etc., where there is no period specifically stated elsewhere within the contract, for example where the Service Manager should reply to the Contractor submitting the particulars of its Equipment design or the name of a proposed Subcontractor. The period of reply would clearly not apply to the contractor's submission of a programme for acceptance or the submission of a quotation for a compensation event as specific time periods are quoted in those provisions.

With regard to dispute resolution, the Adjudicator nominating body is named within this section, normally a professional institution, and then the tribunal is named as either arbitration or legal proceedings.

Finally within this section, any matters to be included in the Early Warning Register will be identified.

Section 2: Contractor's main responsibilities

This section provides for entries in respect of the frequency of the Contractor's forecasts of Defined Cost under Option C and E.

Section 3: Time

The starting date, service period, and the frequency of requirement for the contractor to submit revised plans and Task Order programmes are inserted within this section.

Section 4: Quality management

The period for the contractor to submit a quality policy statement and quality plan are stated here.

Note that there is no defects date or defects certificate within the TSC, though clients may consider amending the Contract to make provision for these, for example stating the number of weeks after the Service Period has ended. This would then identify the period the contractor is initially liable for correcting defects in any construction works it has undertaken itself, for example through task orders.

Clients could also consider inserting defect correction periods, these being the period within which the contractor must correct each notified Defect, failing which the Service Manager assesses the cost to the client of having the Defect corrected by Others.

The Contract Data could then provide for multiple entries for defects correction periods, to be inserted if required, which then allows for different types of defect to have different correction periods dependent on the urgency to have them corrected, for example:

- Mechanical and electrical works:24 hours
- Drainage: 48 hours
- Finishings: 7 days

The contract includes provisions for response times to specified events through secondary option X17, Low service Damages. This provision would be used for incentivizing the Contractor to respond to problems in infrastructure it has not provided itself.

Clearly any Defect that has a potential effect on health and safety should be corrected as soon as possible.

Section 5: Payment

Again, as the NEC4 contracts are intended for use worldwide, the currency of the contract is entered.

The assessment interval is also entered, which again is often expressed as "one calendar month" rather than in "weeks". The Service Manager decides the first assessment date to suit the parties and following assessments are carried out within the assessment intervals.

If Option Y(UK)2 is not used, the period for payment is three weeks from the assessment date (Clause 51.2), though this can be amended in this section, and inserting interest rates for late payment.

The default period for certifying a final assessment is 13 weeks, but this may be amended here.

Where Option C is used the share range(s) and contractor share percentage(s) are also inserted here.

Finally, where Option C or E is chosen, the published exchange rates may be inserted.

Section 6: Compensation events

The default value engineering percentage is 50 per cent, though this can be changed here.

Additional compensation events may also be inserted here.

Section 7: Use of Equipment, Plant and Materials

This section is not included within the Contract Data as no entries are required.

Section 8: Liabilities and insurance

If there are additional client's liabilities (Clause 80.1) these can be listed.

The minimum limit of indemnity for third party public liability and the contractor's liability for its own employees are stated within this section

There are then optional statements regarding insurance.

First, if the client is to provide plant and materials there is provision for insurance of the works to include any loss or damage of such plant and materials.

The contractor provides the insurances stated within the Insurance Table, except any insurances which the client is to provide which are stated in Contract Data Part 1, which lists what the insurance is to cover, the amount of cover and the deductibles (otherwise known as "excesses"), the amounts to be paid by the insuring party, often before the insurance company pays.

RESOLVING AND AVOIDING DISPUTES

First, there is a choice of tribunal which is normally litigation (the Courts) or arbitration.

If the tribunal has been identified as arbitration, the client must identify the arbitration procedure, the place where any arbitration is to be held, who will choose an arbitrator if either the parties cannot agree a choice, or if the arbitration procedure does not state who selects an arbitrator.

The Senior Representatives are also named, also the Adjudicator and/or the Adjudicator nominating body.

Finally, dependent on which Secondary Options are selected, the Client enters details for the relevant entry in the Contract Data.

11.14   Inviting tenders

Each tenderer should be given a complete set of contract documents for the full tendering period. All sets should be identical.

If additional information is issued during the tender period, each such issue should be numbered sequentially and dated, and reach all tenderers in time for them to act on the information provided.

11.15   Submission of tenders

It is critical that tenderers are given sufficient time in which to prepare and submit their tenders.

The time and process (usually online through a web portal) needs to be clearly stated. It should also be made clear to all tenderers that late tender returns are invalid and should not be considered, unless all are given a further opportunity to re-tender. Preferably, late tenders should not be opened.

The tender documentation should also make clear the period over which a tender may remain open for acceptance. This would normally be at least three months, giving the client time to assess the tenders and make the appropriate award.

11.16   Opening tenders

Tenders should not be opened until after the date and time for submission to avoid any suspicion of collusion. No tenderer should be allowed to alter the terms of a tender after the closing time.

After opening the tenders and supporting documents should be evaluated. Each tender is confidential and no details should be released to any other tenderer or to anyone with a financial interest in the tendering.

11.17   Notification of tender results

It is important that all competing contractors be informed of the results of tendering as soon as possible after the tender submission date. This particularly applies to those whose tenders are not successful or which are not being given further consideration as it allows them to concentrate on opportunities with other clients.

Generally, clients' representatives have no unilateral authority to accept tenders, as ultimately the contract is between the client and the contractor, so it is normal to recommend a choice to the client with it making the final decision and informing the tenderers accordingly.

The client's representative should submit a report, which lists the tenders received, comments on them and recommends a particular one for acceptance. The scope and detail of this report will vary according to the circumstances.

The report should present a clear and logically reasoned case for acceptance of the recommended tender. If the inclusion of technical detail is essential, sufficient explanation – in plain language – should also be included.

The following may be included in the report:

- A tabular statement of the salient features of all the tenders received; e.g. the name of the tenderer, the tender sum, whether valid (received by due date, complying with instructions for tendering, etc), and qualifying conditions

- Reference to any discussions that may have taken place with any tenderers regarding the tender
- A concise summary of the findings following the examination and analysis of each tender, reasons for considering any tender invalid, discussion of the plan and methods of execution proposed, and reasons for considering any of these unsatisfactory
- Comments on any rates, which appear exceptionally high or low, and forecasts of the possible effects on the contract price in cases where there are large differences
- A comparison of the recommended tender sum with the cost plan or budget
- Recommendation of the most acceptable tender
- Recommendations for dealing with any qualifications in the recommended tender prior to acceptance

Only after a tender has been accepted, should unsuccessful tenderers be advised of the name and tender price of the successful tenderer.

11.18   Evaluation of tenderers

In evaluating the suitability of the tenderers, the following information may be considered:

- Location
- Corporate structure
- Financial information, bank references, bonding capacity
- Past experience and current service contracts: size, type, performance on schedule and budget
- Current work load
- References
- Work force on contractor payroll vs subcontracting
- Equipment owned vs rentals
- Health and safety issues
- Quality control issues
- Key people to be assigned to the works
- Disputes: record of arbitration/litigation

In addition to checking the information supplied, the client may make independent checks on the contractor's credit standing, visit other clients of the contractor and interview owners and operators, investigate safety and dispute records and review other elements of performance.

11.19   Assessing tenders

By carrying out detailed checks pre-tender, when tenders are received, the evaluation criteria should be very straightforward as all the tenders should be capable of being accepted, so it will normally just include any further information gained from the bid itself:

- Price
- Contractor's ability to achieve quality standards
- The Contractor's proposed team and team structure

There has been a trend in the past 30 years for tenders to be accepted on the basis of quality elements of the bid, rather than on price alone.

REMEMBER – THE CHEAPEST PRICE DOES NOT ALWAYS MEAN THE CHEAPEST COST!

A detailed tender analysis should be prepared showing details of each of the tenders submitted including:

- Price submitted
- Any exclusions or qualifications to the price submitted
- Any arithmetical errors
- Any technical issues such as design requirements
- Compliance between Client's Scope and Contractor's Scope for its plan
- The fee percentage which will apply to compensation events, task orders and – in Options C and E – to routine monthly payments

There should be clear guidance on how to deal with late, incomplete or qualified bids.

11.20   Comparison of tenders

Contractors' tenders are not always straightforward to compare. Some do not price all the items in the enquiry; often there are mistakes and some contractors price net while others offer discounts, typically 2.5 per cent.

Therefore, a register or spreadsheet is a useful device to enable quotations to be matched, discrepancies identified and discounts adjusted.

A typical checking procedure when comparing tenders should include consideration of the following:

- Does the service described in the tender comply with the Scope?
- Have all items have been priced and if not are they included in other rates?
- Are unit rates consistent throughout the tender?
- Check that the tender does not form a counter-offer and that the contractor has accepted the terms and conditions of the enquiry.

All tenders should be checked arithmetically and compared with each other and also with the cost plan for the contract in case of any errors in either. It is good practice to query any tender that is unreasonably low, the tenderer may then be invited to confirm its tender price or withdraw the tender.

Tender sums from equally competent contractors may vary considerably, not so much because the routine operations are differently priced, but because of different service delivery philosophies and techniques and because of differing values attached to risks and to the methods of dealing with them. Profit margins included in tenders may also differ.

Any obvious mistake should be drawn to the attention of the tenderer. If the mistake results from an apparent misunderstanding of the documents and is common to several tenders, it may be necessary to invite new tenders. The tenderer should be given the opportunity to either confirm or withdraw its tender.

Usually it is considered inappropriate to allow a tenderer to correct any mistake (e.g. in a rate, its extension, or its summation), because the range of tenders is probably already known and it might give this tenderer an unfair advantage over other tenderers. However, some may wish to allow correction anyway.

If supplementary information that was called for is not included, this should be obtained as soon as possible for tenders that are to be seriously considered.

11.21 Qualifications and ambiguities

It is often stipulated that no qualifications to the tender will be allowed; e.g. alternative proposals etc. Indeed, if tenders are to have a common basis for comparison, it is essential to stipulate that every tenderer submits a tender in the form requested. So long as this is done, a tenderer should not be penalised for offering alternative proposals.

11.22 Awarding the Contract

Once the tenders have been assessed and a decision made as to which tender is successful and the contract awarded, it is essential that the contract be awarded as soon as practicable, and also that unsuccessful tenderers be advised accordingly.

At this stage, it is useful to follow a checklist approach to ensure that all the pre-construction issues are covered.

These will include:

- Confirm order or issue letter of intent
- Prepare contracts for signing
- Start any people transfer processes
- Confirm requirements for insurances
- Agree or obtain client/client representative approval of subcontracts or packages where there is a contractual requirement to do so
- Agree the facilities to be provided by the contractor, for example, site offices, stores, offloading, scaffolding, removing rubbish, etc.
- Agree dates for interim certificates and procedure for requesting payment
- Agree notification requirements and any pro formas for instructions, early warnings, compensation events, etc.
- Sign contract

There is a suggested Form of Agreement for completion and signature within the NEC 4 publication, "Preparing a Term Service Contract"

If they choose to do so, the parties may write their own, and in that respect must be very cautious that all the documents that are intended to form the contract are included and stated on the Form of Agreement.

Following the report on tenders the next move is with the client. Decisions taken on any matters which are not included in the contract documents, or which vary them, must be carefully recorded by means of an exchange of letters.

In order to satisfy the common law test of offer and acceptance, the acceptance of a tender must be unconditional. Thus a letter of acceptance is not the place to introduce additional conditions like "your tender is accepted subject to our obtaining the necessary finance within one month". Such new conditions would be a counter-offer, cancelling the original offer and in turn requiring acceptance by the tenderer. While some negotiation between the parties is not unusual, eventually an unconditional acceptance of a current offer (or counter-offer) is needed.

11.23 Formalising the Contract

The signing of a formal agreement only adds additional evidence to the contract, but is still desirable, particularly if there have been complex negotiations. It is important to ensure that the signed documents incorporate all the documentation on which tenders and the final offer and acceptance are based. At the time of signing of the agreement, the parties, together with any amendments, which may have been made to printed documents, should initial every page of all such other documents.

When arranging the signing of contract documents, a check should be made that the signatories have legal authority to make contracts on behalf of their organisation, and that the formalities required by the organisation's rules are followed. This arises particularly with companies, incorporated societies or statutory bodies; typically they may require the placing of the organisation's seal and two authorised signatures.

11.24   Letter of acceptance of tender

This is a letter from the Client to the Contractor accepting the offer. If bargaining between principal and contractor takes place before a tender is finally accepted, copies of any other letters giving evidence of the conditions agreed to should also be incorporated in the documents. In order to form a contract, there must be a well-defined offer to which an unconditional acceptance has been given.

12 Comparison of the Term Service Contract (TSC) with the Professional Service Contract (PSC) and the Facilities Management Contract (FMC)

It is not proposed to carry out a detailed analysis of the Professional Service Contract (PSC) and the Facilities Management Contract (FMC) within this Chapter, but to compare the two contracts with the Term Service Contract (TSC) which is the subject of the book, and which has been explained in detail within earlier Chapters, in order to highlight the most important differences in their use, structure and content.

While at first the three contracts appear to be very similar in that they are all for providing a service, and they all share NEC features such as Main and Secondary Options, plans and programmes, early warnings and compensation events, they are used for different types of services and therefore have some notable differences.

Let us consider each in turn:

The Term Service Contract (TSC)

The TSC is used to appoint a Contractor to provide a wide range of services over a specific period of time, which can span over a number of years (the "Service Period") and in various locations which can be over a wide geographical area (the "Affected Property").

Examples of the use of the TSC in the construction sector include:

- maintenance of highways for a local authority
- regular upkeep of public parks and landscaped areas in a city
- maintaining mechanical and electrical systems within buildings
- maintaining lifts in a group of hospitals

The authors have had considerable experience of advising and training practitioners using the TSC for these types of services.

Also as stated in the Introduction to this volume, the TSC is used, for non-construction services, for example:

- cleaning of streets in a town centre
- refuse collection and disposal
- provision of community ambulance services for a group of hospitals

The NEC4 contracts used for a term service are:

- Term Service Contract (TSC)
- Term Service Subcontract (TSS)
- Term Service Short Contract (TSSC)

DOI: 10.1201/9781003463771-13

The Professional Service Contract (PSC)

The PSC is used to appoint a Contractor to provide a specific service, normally over a shorter period than the PSC or FMC, and also in connection with a specific project in a specific location.

Examples of the use of the PSC include:

- architectural/structural design of a project
- a technical survey of a piece of land to compile a flood risk assessment
- a commercial service such as quantity surveying in relation to a project

The NEC4 contracts used for a professional service are:

- Professional Service Contract (PSC)
- Professional Service Subcontract (PSS)
- Professional Service Short Contract (PSSC)

The Facilities Management Contract (FMC)

The FMC is used to appoint a Service Provider to manage and provide a facility management service over a specific period of time, which can span over a number of years (the "Service Period") and in various locations which can be over a wide geographical area (the "Affected Property").

So, let us start by defining facilities management (FM):

The International Organisation for Standardisation defines facilities management (FM) as the "*organizational function which integrates people, place and process within the built environment with the purpose of improving the quality of life of people and the productivity of the core business*".

So facilities management services will normally apply to providing a service within a building, or set of buildings i.e. someone's workplace.

So, as an example an insurance company's core business is providing services in the field of insurance, but in order to focus and be efficient in that business, the insurance company may employ a facilities management company to take care of non-core activities such as maintenance of their buildings, window and façade cleaning, security, catering and other services.

The Facilities Management Contract was drafted jointly with the Institute of Workplace and Facilities Management specifically for the procurement of facilities management services, helping to identify best practice in the sector and establish the way forward.

Note that prior to the FMC being launched in June 2021, the NEC3 and NEC4 TSC was regularly used for facilities management contracts, and the authors have significant experience in training companies in the use of the TSC for that purpose, although it must be said that the FMC is significantly more fitting for this purpose.

Examples of the use of the FMC include:

- Building maintenance including mechanical and electrical systems
- Daily office cleaning
- Periodic cleaning of windows and the external façade of a building

- Security services
- Catering services

The NEC4 contracts used for a facilities management service are:

- Facilities Management Contract (FMC)
- Facilities Management Subcontract (FMS)
- Facilities Management Short Contract (FMSC)
- Facilities Management Short Subcontract (FMSS)

A comparison between the TSC, the PSC and the FMC

Refer to Figure 12.1 at the end of this chapter comparing the contents of the TSC, PSC and FMC.

(i) Core clauses

The titles to the core clauses are almost the same, apart from Clause 8 which under the TSC and FMC is titled "Use of equipment. Plant and Materials" while under the PSC it is titled "Rights to Material".

Note that the contents within the core clauses are different across the TSC, PSC and FMC, some of which will be explained below.

Note also that the provider of the service has a different title across the three contracts:

- TSC – the "Contractor"
- PSC – the "Consultant"
- FMC – the "Service Provider"

Note that the Client's representative in all three contracts is the "Service Manager".

(ii) Main Options

The Main Options within the TSC have been discussed in detail in the Introduction to this book, so it is not intended to detail them again, merely to highlight the most important differences between the TSC, the PSC and the FMC.

The TSC and FMC have the same Main options:

- Option A: Priced contract with Price List
- Option C: Target contract with Price List
- Option E: Cost reimbursable contract

It is not proposed to examine each of these Main Options in detail, but note that under Option A, the Contractor (TSC) or the Service Provider (FMC) is paid for completed items in the Price List, plus where there is a quantity for an item, the quantity carried out to date at the rates in the Price List.

The PSC has the following Main Options:

- Option A: Priced contract with activity schedule
- Option C: Target contract
- Option E: Cost reimbursable contract

Again, it is not proposed to examine each of these options in detail, but note that under Option A, the Consultant is only paid for completed activities within the activity schedule.

(iii)   Secondary Options

Again, the Secondary Options within the TSC have been discussed in detail in the Introduction to this book, so it is not intended to detail them again, merely to highlight the most important differences between the TSC, the PSC and the FMC.

At first glance, the Secondary Options for the TSC, PSC and FMC appear similar, but there are some exceptions.

Additional Secondary Options within the PSC, but not within the TSC or FMC:

• Option X5: Sectional completion

The service may be divided into pre-defined sections.

• Option X6: Bonus for early completion

The Client may pay a bonus to the Consultant for early completion of the service.

• Option X6: Delay damages

The Consultant may pay delay damages to the Client for late completion of the service.

Additional Secondary Options within the FMC:

Note that Options X4 (Ultimate Holding Company Guarantee) and X13 (Performance Bond) within the TSC and the PSC are combined as Option X4 (Performance Guarantee) within the FMC.

• Option X15: The Service Provider's design

This option requires the Service Provider to submit particulars of any design it is undertaking, the design being carried out using reasonable skill and care, the period for retention of documents in case of claims, and also professional indemnity insurance requirements.

• Option X27: Project Orders

In both the TSC and FMC, there is a process for ordering new work that has a level of complexity and risk which requires a programme for its execution, and which may include changes or additions to the Scope and Price List.

In the TSC such work is instructed under Task Orders which form part of the core clauses and are therefore always available for use.

In the FMC such work is instructed under Secondary Option X27 Project Orders, which will need to be selected for inclusion if required.

• Option X28: Change of control

As the service Period for a FMC can be several years, there is a possibility that there may be a change of control of the Service Provider by an individual, a company or a partnership.

It gives the Client the option of termination of the FMC if the individual, company or partnership is not financially strong enough, it contravenes ethical principles, or there is an unresolvable conflict of interest.

Additional Secondary Options within the PSC and FMC:

• Option X9: Transfer of rights

This option means that the Client owns the Consultant's (PSC) or the Service Provider's (FMC) rights over material prepared for its design, and obtains equivalent rights from a Subcontractor.

(iv) Other features of the PSC and FMC

The PSC

The PSC provides all of the features of a consultancy contract, but it is drafted in a similar way to the other NEC contracts, so it includes requirements for early warnings, a detailed programme submitted by the Consultant to the Service Manager for acceptance, and also compensation events which include changes to the Prices and any delay to the Completion Date or Key Dates.

Note that with compensation events under the TSC and the FMC, the Prices may be changed, but the Service Period is not changed, with the TSC a Task Completion Date may be delayed and under the FMC the Performance Table may be changed.

The FMC

There are a number of features of the FMC that need some explanation:

(i) Service Failures

The FMC does not use the term "Defect", it calls it a "Service Failure" which is a term more relevant to the provision of a service. The Scope may include a Service Level Agreement, so the service provided has fallen short of that.

(ii) Service Orders

With the FMC additional work can be called off using Service Orders which are an instruction to carry out work identified in the Scope on an "as instructed" basis.

The Service Manager may issue a Service Order to the Service Provider, which includes a detailed description, a time period, and other information. The Service Order is priced using rates and Prices in the Price List.

The form of the Service Order is detailed in the Scope, and will include any processes to be followed in the issue and execution of Service Orders and any constraints (for example, on the value of work that can be done or the procedures for getting approval before doing the work).

The Service Order process in the FMC creates a simpler and more structured process than Task Orders under the TSC.

(iii) Mobilisation plan

The Service Provider submits a mobilisation plan to the Service Manager for acceptance, the Service Manager replying with acceptance or non-acceptance within one week.

(iv) Demobilisation plan

The Service Provider submits a demobilisation plan to the Service Manager for acceptance, the Service Manager replying with acceptance or non-acceptance within one week.

 (v) Performance measurements

Throughout the Service Period the Service Provider is required to report its performance against the targets in the Performance Table.

The Performance Table sets the targets the Service Provider is to achieve in Providing the Service and the adjustments to payment if a measured performance is higher, the same or lower than its target.

This replaces Secondary Option X17 (Low Performance Damages) and X20 (Key Performance Indicators) within the TSC.

Term Service Contract	*Professional Service Contract*	*Facilities Management Contract*
Core Clauses	**Core Clauses**	**Core Clauses**
1. General	1. General	1. General
2. The Contractor's Main Responsibilities	2. The Consultant's Main Responsibilities	2. The Service Provider's Main Responsibilities
3. Time	3. Time	3. Time
4. Quality Management	4. Quality Management	4. Quality Management
5. Payment	5. Payment	5. Payment
6. Compensation Events	6. Compensation Events	6. Compensation Events
7. Use of equipment, Plant and Materials	7. Rights to Material	7. Use of equipment, Plant and Materials
8. Liabilities and Insurance	8. Liabilities and Insurance	8. Liabilities and Insurance
9. Termination	9. Termination	9. Termination
Main Options	**Main Options**	**Main Options**
Option A : Priced Contract with Price List	Option A : Priced Contract with activity schedule	Option A : Priced Contract with Price List
Option C : Target Contract with Price List	Option C : Target Contract	Option C : Target Contract with Price List
Option E : Cost Reimbursable Contract	Option E : Cost Reimbursable Contract	Option E : Cost Reimbursable Contract
Secondary Options	**Secondary Options**	
Option X1 : Price adjustment for inflation	Option X1 : Price adjustment for inflation	Option X1 : Price adjustment for inflation
Option X2 : Changes in the law	Option X2 : Changes in the law	Option X2 : Changes in the law
Option X3 : Multiple currencies	Option X3 : Multiple currencies	Option X3 : Multiple currencies
Option X4 : Ultimate holding company guarantee	Option X4 : Ultimate holding company guarantee	Option X4 : Performance guarantee
	Option X5 : Sectional Completion	
	Option X6 : Bonus for early Completion	
	Option X7 : Delay damages	
Option X8 : Undertakings to the *Client* or Others	Option X8 : Undertakings to Others	Option X8 : Undertakings to the *Client* or Others
	Option X9 : Transfer of rights	Option X9 : Transfer of rights
Option X10 : Information modelling	Option X10 : Information modelling	Option X10 : Information modelling
Option X11 : Termination by the *Client*	Option X11 : Termination by the *Client*	Option X11 : Termination by the *Client*
Option X12 : Multiparty collaboration	Option X12 : Multiparty collaboration	Option X12 : Multiparty collaboration
Option X13 : Performance bond	Option X13 : Performance bond	Option X15 : The Service Provider's Design
Option X17 : Low service damages		
Option X18 : Limitation of liability	Option X18 : Limitation of liability	Option X18 : Limitation of liability
Option X19 : Termination by either Party		Option X19 : Termination by either Party
Option X20 : Key Performance Indicators	Option X20 : Key Performance Indicators	

(*Continued*)

(Continued)

Term Service Contract	Professional Service Contract	Facilities Management Contract
Option X21 : Whole life cost		Option X21 : Whole life cost
Option X23 : Extending the Service Period		Option X23 : Extending the Service Period
Option X24 : The *accounting periods*		Option X24 : The *accounting periods*
		Option X27 : Project Orders
		Option X28 : Change of Control
Option X29 : Climate Change	Option X29 : Climate Change	Option X29 : Climate Change
Option Y(UK)1 : Project Bank Account	Option Y(UK)1 : Project Bank Account	Option Y(UK)1 : Project Bank Account
Option Y(UK)2 : The HGCR Act 1996	Option Y(UK)2 : The HGCR Act 1996	Option Y(UK)2 : The HGCR Act 1996
Option Y(UK)3 : Contracts (Rights of Third Parties) Act 1999	Option Y(UK)3 : Contracts (Rights of Third Parties) Act 1999	Option Y(UK)3 : Contracts (Rights of Third Parties) Act 1999
Option Z : *Additional conditions of contract*	Option Z : *Additional conditions of contract*	Option Z : *Additional conditions of contract*

Index

Note: All items within this index refer to the NEC4 TSC (Term Service Contract), but items marked "(FMC)" refer to the NEC4 Facilities Management Contract specifically referred to in the comparison with the TSC and the PSC (Professional Service Contract) within Chapter 12.